埃琳娜·曼费迪尼作品解析（上）

LANDSCAPES

水 平

李宁　[意]埃琳娜·曼费迪尼　著

江苏凤凰科学技术出版社

图书在版编目（CIP）数据

埃琳娜·曼费迪尼作品解析 / 李宁，（意）埃琳娜·
曼费迪尼著 . -- 南京：江苏凤凰科学技术出版社，2020.1
 ISBN 978-7-5713-0654-0

Ⅰ . ①埃… Ⅱ . ①李… ②埃… Ⅲ . ①建筑设计－作
品集－意大利－现代 Ⅳ . ① TU206

中国版本图书馆 CIP 数据核字（2019）第 248258 号

埃琳娜·曼费迪尼作品解析

著　　　者	李　宁　[意]埃琳娜·曼费迪尼
项 目 策 划	凤凰空间/曹　蕾
责 任 编 辑	刘屹立　赵　研
特 约 编 辑	曹　蕾

出 版 发 行	江苏凤凰科学技术出版社
出 版 社 地 址	南京市湖南路1号A楼，邮编：210009
出 版 社 网 址	http://www.pspress.cn
总　经　销	天津凤凰空间文化传媒有限公司
总 经 销 网 址	http://www.ifengspace.cn
印　　　刷	天津久佳雅创印刷有限公司

开　　　本	787 mm×1 092 mm　1/16
印　　　张	20
版　　　次	2020年1月第1版
印　　　次	2020年1月第1次印刷

标 准 书 号	ISBN 978-7-5713-0654-0
定　　　价	398.00元（上、下）（精）

图书如有印装质量问题，可随时向销售部调换（电话：022-87893668）。

CONTENTS
目录

4	FOREWORD 前言
6	ACKNOWLEDGEMENTS 致谢
8	INTRODUCTION 简介
12	NOTE ON CODING AND GEOMETRICAL COMPOSITION 浅论编码和几何构图
16	LANE CRAWFORD 连卡佛
44	URBAN FABRIC RUGS 城市肌理地毯
54	FRAMED SKYLINES 天际轮廓线
60	LOO WITH A VIEW 观景洗手间
68	MASSIVE PROJECTIONS 巨幕投影
74	FRESCO 壁画
80	FABULATIONS 虚构情节
88	EYE CANDY 秀色可餐
96	TEMPERA 彩画
112	INVERTED LANDSCAPES 倒置景观
128	LIVING PICTURE 生活图片
142	IMAGE ANNOTATION 图片注释

FOREWORD

BY XU WEIGUO

I met Elena Manferdini in 2006, when I was one of the curators for the second Architecture Biennial in Beijing (ABB), in which she was invited to participate in the exhibition "Emerging Talents, Emerging Technologies". In addition to exhibiting her architectural designs, she was asked to create and build the West Coast Pavilion. As the first Chinese building made with CNC technology, the West Coast Pavilion opened the door to digital construction. At its opening, the plaza in front of The China Millennium Monument attracted a large crowd of onlookers: people were drawn to this vision of architectural detailing and articulation which was simultaneously digital and organic in its appearance. The Pavilion offered a new kind of architectural form, and with it an experience like had never been seen before.

The integration of digital technology and architecture in China has developed rapidly in depth and breadth. It has not only improved the efficiency and quality of existing architectural design, but also radically changed the way architects think. New smart construction technologies are being developed and will quickly be popularized, altering the field even further as they make previously unimaginable architectural designs possible. It should be noted that Elena, as a Western scholar, has played an important role in this digitalization of Chinese architecture through her articles, projects, lectures, exhibitions, etc. She has and continues to influence the younger generation of students and architects through her numerous contributions to the prosperity of the industry.

This book collects Elena's digital design research from the last six years, published as two volumes, Landscapes and Portraits. The former is composed of "pictures" and the latter of "drawings", both based on digital technology. Landscapes uses 3D scans of natural elements, such as flora and fauna, to create processed representations which oscillate between the real and the imaginative in order to present a new, tactile composite environment. Portraits is composed of "code-generated graphics". None of the drawings are like the usual architectural drawings—typically a subtraction process from a three-dimensional object to a two-dimensional plane—they are more like a proportionally scaled-down representation of a building. The book is the result of Elena's unique and in-depth thinking on the combination of digital technology and architectural design. It leads us to re-examine our techniques and theories of architecture through the lens of these newfound contemporary methodologies.

前言

徐卫国

认识埃琳娜·曼费迪尼（Elena Manferdini）是在2006年，这年我作为策展人之一，第二次策展北京中国建筑艺术双年展，并邀请埃琳娜参加了在中华世纪坛举办的"涌现"建筑展。埃琳娜不仅展出了她的建筑设计作品，还应邀创作并建造了"西海岸展亭"。更有趣的是，展览开幕式还邀请埃琳娜设计制作了一套基于数字技术设计加工的时装，并请当时北京最好的模特进行了表演，参加开幕式的近千人观看了这个表演。"西海岸展亭"是一个不得了的作品，可以说它是中国第一个用数控技术加工建造的构筑物，开启了数字建造的历史。当时在中华世纪坛前广场上确实吸引了络绎不绝的人群，引起了不小的轰动，让人们亲眼看到了一种新的建造方式，给人们带来一种前所未见的形式及体验。

众所周知，从2004年到2019年的15年间，中国数字技术与建筑设计的结合在深度及广度上得到了迅速的发展，它不仅提高了现有建筑设计的效率和质量，并且正在改变建筑师的设计思考方式以及设计工作过程，同时也在改变实际项目的建设组织方式及实施过程。特别是由于人口红利的消失、劳动力价格的迅速攀升，大量建设工程面临工人短缺的困境，这样一来，用机器替代人力将成为必由之路，因此，数字设计及数字建造正在使建筑工业升级换代，新的智能建造技术正在酝酿并将迅速得到普及。应该说，埃琳娜作为西方学者，通过文章、作品、演讲、展览等方式，对中国建筑的这一数字化过程产生了重要的影响，她影响了年轻一代学生以及建筑师，进而推动了这一发展进程。

本书收集了埃琳娜近6年的数字设计研究成果，分为"水平"与"竖直"两册出版，前者是"图片"，后者是"图纸"，两者都基于数字技术产生。前者是"通过3D扫描采集的自然元素，如植物群和动物群，初始状态是三维数字点云，将它们进行处理形成这些图片"，呈现出崭新的充满触感的合成环境；后者是"通过编写代码生成的图形，没有一张像通常建筑图一样是从三维到二维平面的减法过程获得的结果，它们就像按比例缩小的建筑物再现"。这是埃琳娜对数字技术与建筑设计相结合这一问题进行独特而深度思考的结果，它让我们重新审视在数字技术条件下建筑设计的"表现"方法与理论。我想无论西方学者还是中国学者都一定会对这本书产生强烈的反响。

ACKNOWLEDGEMENTS

BY ELENA MANFERDINI

I would like to thank the institutions that supported the realization of this publication: Beijing University of Technology, and especially Dr. Li Ning, as well as Tsinghua University as represented by Professor Xu Weiguo. Thank you for giving me the opportunity to look back at my professional work and see it through a new lens. Every page of these two volumes is a journey that brings clarity and purpose to the creative impetus of the work.

I would also like to thank the Southern California Institute of Architecture for being a place of academic and professional growth. During the past 16 years which I've dedicated to teaching at this institution, SCI-Arc has been an incubator of ideas and experimentation. On a personal level, this school nurtured my creativity, sharpened my critical thinking, and ignited my dreams. I still smile when I take the freeway and cross the city to go to a school full of fearless energy and unbelievable talent.

In particular, my gratitude goes to all of the students and alumni who have been an integral part of the ideation and execution of the work portrayed in this publication. The creative exuberance and technical skills of all the people working in my office during the past six years have been the fuel of the engine. Thank you to Adil Bokhari, Alex Dannecker, Amanda Webber, Andrea Cadioli, Andres Gandara, Ann Gutierrez, Anna Penzo, Asmaa Abu Assaf, Begum Baysun, Jacob Blankenship, Caroline Dieden, Chengxi Hou, Christina Griggs, Christy Coleman, Cindy Yiin, Connor Gravelle, Cristina Toth, Donovan Ballantyne, Dutra Brown, Emre Turan, Erin Templeton, Eugene Kosgoron, Evaline Huang, Farnoosh Rafaie, Feng Wang, Francis Martinez Betances, Ge Wang, Gregory Kokkotis ,Gyasi Williams, Huijin Zheng, Jaegeun Lim, Jasper Gregory, Jean-Pierre Villafañe, Jinsa Yoon, Julia Arnolds, Julia Pike, Kate Gensing, Ke Li, Laura Ferrarello, Lauren Diaz, Leonora Bustamante, Lukasz Blonski, Luis Gerardo Tornel, Macus Hoh, Meenakshi Dravid, Midori Mizuhara, Min-ah Kim, Monica Ximena Gutierrez, Mustafa Kustur, Nichole Tortorici, Nick Rademacher, Omer Pekin, Palak Mandhana, Ronny Eckels, Ross Fernandes, Sara Gaskari, Sean Lee, Sona Gevorkyan, Thomas Cheng, Vanessa Teng, Xuan He, Yu Nong Khew, Zach Hoffmann, Zaid Kashef Alghata and to all those who have been a part of my getting there. Without you the work produced in my office would simply not exist. A special thank you to Shawn Rassekh and Dylan Perkinson for all of their help with the growth of this office during these recent years, and for always having my back when it came to building buildings.

Publishing a book is harder than I thought and more rewarding than I could have ever imagined. A very special thanks goes to the group of contributors who gave words to the work: Hernan Diaz Alonso, Jasmine Benyamin, Gabriel Esquivel, Jeffrey Kipnis, Li Ning, Yael Reisner, and Xu Weiguo — all dear friends and colleagues that have been sources of continuous support, friendship and generosity towards me over the years. Their commitment to architecture is something I have always admired.

My personal gratitude goes to all of the clients and institutions that believed in the projects of Atelier Manferdini and commissioned the work in this publication: Ayala, Art Institute of Chicago, Chicago Architecture Biennial, Design Art Work Fair, Design Matters, Industry Gallery, Infonavit, Istanbul Design Biennale, LA Metro, LACMA, Lane Crawford, La Peer Hotel, Longmont Museum, Los Angeles County Arts Commission, MAK, McGhee James and Joyce, and Urban Fabric Rugs.

I have to end by thanking my family. I'm forever indebted to my husband, Fabio Zangoli. From the beginning of our relationship, you have brought strength and structure to my dreams and never once said it was not possible. Thank you so much, dear. Thank you to my mother and father for taking my calls at all times, and for mastering social media and text messaging in an attempt to stay in contact while I am traveling all around the globe.

This book is dedicated to my son, Tommaso, whose endless curiosity in the world sparks my own.

致谢

埃琳娜·曼费迪尼

我要特别感谢北京工业大学的李宁博士和清华大学的徐卫国教授对此书的出版所给予的支持。感谢给我这个机会，使我可以重新审视自己的专业工作，用新的视角去看待它。这两册书中的每一页都是一段旅程，都是我创作的清晰明确的源动力。

我还要感谢在学术和专业方向上都在不断成长的南加州建筑学院。过去的16年里，我一直执教于南加州建筑学院，这里一直是思想和实践的孵化器。这所学校培养了我的创造力，磨炼了我的批判性思维，点燃了我的梦想。每次当我开下高速，穿过城市来到这所充满无畏精神和惊人才华的学校时，我都是满心欢喜的。

我特别要感谢所有的学生和校友，他们是本书的构思和执行工作中不可或缺的一部分。在过去的6年里，工作室全体员工的创造力和技术技能一直是我们的驱动力。感谢阿迪尔·博卡里，亚历克斯·丹内克，阿曼达·韦伯，安德里亚·卡迪奥利，安德雷斯·甘达拉，安·古铁雷斯，安娜·彭佐，阿斯玛·阿布·阿萨夫，贝古姆·贝松，雅各布·布兰肯希普，卡罗琳·迪登，侯承希，克丽丝蒂·科尔曼，辛迪·伊恩，康纳·格拉弗尔，克里斯蒂娜·托特，多诺万·巴兰蒂娜，杜特拉·布朗，埃姆·图兰，艾琳·邓普顿，尤金·科斯戈隆，埃瓦琳·黄，法努什·拉斐尔，王峰，弗朗西斯·马丁内斯·贝坦西斯，王鸽，格雷戈里·科科蒂斯，吉亚斯·威廉姆斯，郑惠瑾，林杰君，贾斯珀·格雷戈里，让·皮尔法夫，金萨尹，朱莉娅·阿诺德，凯特·根辛，李可，劳拉·费拉雷洛，劳伦·迪亚兹，莱昂诺拉·布斯塔曼特，卢卡斯·布隆斯基，路易斯·杰拉尔多·托内尔，马库斯·霍姆，米纳克什·德拉维德，米多里·瑞祖哈拉，金敏亚，莫妮卡·西梅娜·古铁雷斯，穆斯塔法·库斯托尔，尼古尔·托托里奇，尼克·雷德马赫，奥马尔·佩金，帕拉克·曼达纳，罗尼·埃克尔斯，罗斯·费尔南德斯，莎拉·加斯卡里，肖恩·李，索纳·格沃基安，托马斯·程，凡妮莎·滕，贺璇，丘语农，扎克·霍夫曼，扎伊德·卡西夫·阿尔加塔。感谢所有为本书成功出版做出贡献的人，没有你们，我的工作也无法开展。特别感谢肖恩·拉塞克和迪伦·帕金森，感谢他们近年来对工作室的壮大所给予的一切帮助，同时感谢他们在建筑方面对我一直以来的支持。

出版一本书比我想象的难却更有价值。特别感谢那些工作谏言的贡献者：埃尔南·迪亚兹·阿隆索，杰思敏·本雅明，加布里埃尔·埃斯奎维尔，杰弗雷·基普尼斯，李宁，雅尔·雷斯纳，徐卫国，感谢所有亲爱的朋友和同事不断的支持，感谢她们的友好和慷慨，敬佩他们对建筑的执着。

我本人再次感谢所有支持曼费迪尼事务所项目并授权出版相关作品的客户和机构：阿亚拉，芝加哥艺术学院，芝加哥建筑双年展，设计艺术作品展，设计画廊，工业画廊，因福那维，伊斯坦布尔设计双年展，洛杉矶地铁，洛杉矶艺术博物馆，连卡佛，拉佩尔酒店，朗蒙特博物馆，洛杉矶艺术委员会，MAK，麦吉詹姆斯和乔伊斯以及城市织物毯。

最后，我要感谢我的家人，感谢我的丈夫法比奥·萨格里，从我们在一起到现在，你持续支持我并为我的梦想赋予力量，从未说过泄气的话，非常感谢，亲爱的！感谢我的父母一直以来的关心，感谢他们学会使用社交媒体和发短信，让我在环游世界的时候仍能与他们保持联系。

这本书献给我美丽的儿子托马索，你对世界无尽的好奇一直激发着我的灵感。

INTRODUCTION

by Elena Manferdini

"The Old Masters had sensed that it was necessary to preserve what is called the integrity of the picture plane: that is, to signify the enduring presence of flatness underneath and above the most vivid illusion of three-dimensional space."

Clement Greenberg[1]

This monograph collects the work created by Atelier Manferdini during the last six years of design activity. The projects are not organized in chronological order, but are subdivided into two large families of explorations, Landscapes and Portraits. The titles of the two volumes allude to two distinct picture plane orientations: horizontal and vertical. In art and descriptive geometry, one refers to the picture plane as an abstract plane located between the eye and the object of interest. Traditionally, the picture plane is perpendicular to the axis that comes straight out of an imaginary eye.

When the picture plane displays a portrait orientation, the top usually corresponds to the viewer's head, while its lower edge gravitates to the feet. A portrait orientation represents a world space, as seen by the upright human posture of the artist that is staring at it. The orientation not only represents an act of vision by the author, it also orients the audience. It imposes a way of seeing. On the other hand, the landscape orientation challenges the dimensional bounding box of the erect human posture and becomes an analogue of a visual experience. Landscapes, unlike portraits, represent not a way of seeing, but more so a way of being in the world. While the portrait orientation evokes the presence and the ergonomics of the human protagonist, the landscape is projecting outside of the eye of the beholder, into the proportion of the larger-than-human vista.

The two titles, Portraits and Landscapes, divide the recent work of Atelier Manferdini, and address two distinct natures of the psyche of the image. Each title embodies a special mode of imaginative confrontation. To be clear, landscapes and portraits are not only about how one orients an image, but also how that image is produced and how it engages the beholders. In addition, the rotation of the picture plane from portrait to landscape evokes a shift in the subject matter from abstract geometries to literal figurations. While the Portraits volume collects what we traditionally refer to as architectural drawings, meaning geometrical armatures that start with an abstract set of rules (non-image) in order to generate an observable object, effect, or image; the Landscapes volume collects pictures that work in the opposite directions: they take as point of departure observable objects and arrive to abstraction.

In particular, the pictures collected in the first volume of the publication, Landscapes, are radically different from the traditional transparent projection planes of a horizon as seen from the human eye. These landscapes are produced in an orthogonal plan; they have no camera eye, no privileged view, they have no perspective. They are not associated with a defined status or scale; they have no depth. Their flatness brings together illustration, animation, photography, and graphic design in a fine-art space which collapses their hierarchical distinctions. Their subject matter is usually photorealistic and the original materials for each depiction are 3D scanned acquisitions of natural elements such as flora and fauna. The scans result in three dimensional digital clouds of points and textures which were then digitally manipulated as though they were made of paint.

Such pictures are forms of representation intended for an audience and not as an exercise of personal style. Their success is measured by the degree to which they generate a theatrical engagement with the wider public. Realism and familiarity are not viewed as shortcomings, but rather as crucial ingredients for the success of a landscape. The very existence of the beholder (audience) emerged in art and theater as a fundamental problem of a stage. The theatrical is about the possibility to enter that composition.

---

1 "Modernist Painting, Art and Literature", Spring 1965

The correlation between photorealistic representation and public reaction has become, in my opinion (once again), of interest to architects. These images embody a new generation of synthetic environments in which special attention is paid to the literal reproduction of matter and its tactile effects, where familiarity is the result of multiple mutations from reality and becomes an indispensable ingredient to establish a connection with the audience. These landscape pictures are therefore potentially the ground for a new trope of digital sensibility where real and fantastic, familiar and unexpected, and still and living are increasingly blurred.

Landscapes addresses the status of photorealism in architectural representation, claiming a new pictorial space for architecture between fine arts and digital media. The first volume tackles the problem of depictions as specific kinds of images which are able to embrace issues of figuration and narrative. These pictures are embedded with optical effects and layered with material finishes. The work claims that surfaces now have an unprecedented ability to create sensorial effect, cultural associations, and to re-open a discussion on the role of fantasy in architecture. On a personal note, the work is a silent protest against renderings as we know them today. This work aspires to prove that photorealism, so dear to architects of this digital era, has the ability to move beyond commercial applications and become instead a powerful and evocative creative tool.

The second volume, Portraits, is a collection of architectural drawings, mostly building elevations. The work explores the potential of intricate scripted line work in depicting building facades. The collection exists simultaneously as architectural research and as autonomous artwork. These drawings can be understood as scaled down reproductions of buildings, and at the same time as full-scale printed artifacts.

The reason for compiling a suite of digital sketches is rooted in the belief, that for the past twenty years, computers have been able to produce new geometrical forms that can no longer be understood solely through conventional representational devices like elevations, plans and sections. Buildings nowadays are multi-directional volumes that cannot be precisely oriented; and as a consequence, canonical drawing conventions have become ambiguous — some might even argue they are outdated.

Portraits explores how drawings, and in particular building elevations, could be updated to become once again a creative instrument rather than a reductive one. Specifically, none of these drawings are the result of a subtractive process from the third dimension to the flat plane (as architectural drawings usually are), but they are instead the starting point for endless three-dimensional interpolations. These drawings are agile and are cut, pasted, projected, extruded and folded in various configurations in an abstract and forgiving flat plane. The work exploits the ability of bi-dimensional drawings to imply multiple simultaneous readings and numerous physically three-dimensional manifestations. They are a reflection on our contemporary digital visualization tools, and how their multidimensional hybrid nature is able to accelerate our creativity.

Most of these elevations were produced in rapid succession, one after the other. The drawings have been scripted by code and then were composed into sketches. They remain, intentionally, somewhat abstract; their realm is that of ink, and their role is suggestive of a sketch rather than a true elevation. The work reveals a slavery to a recursive digital era and suggests how that sense of order transfers through fabrication to various scales of design. The projects are characterized by a maniacal attention to geometrical repetitive systems and claim a territory that connects aggregational logics, patterns, and grids with tectonic solutions. Portraits supports the continued practice of thinking through drawings and suggests that these didactic instruments are not outdated, but can be updated and used to trigger our contemporary imagination. The collection of Portraits conveys a sense of immanence—which directs the mind to a reality not yet in existence.

Historically, architectural visualizations have always been the product of an inherent dichotomy because they are usually conceived as the expressions of 'artistic' ability combined with 'technical' savoir-faire. They visualize the tension between intuition and reflection, and between concreteness and abstraction. In other words, they oscillate between being a way to imagine a plausible reality, or simply a means to an end.

With the digital revolution, the dual nature of the drawing vs pictures became even more polarized. On one hand there are technical drawings, and on the other photorealistic renderings. Each giving rise to a perceived division between visualizations made for a professional audience and those created for the general public. With the digital turn, the pictorial connotations of the drawing run into a common risk of shifting from the expression of a specific creativity—as it has been in the past—to the commercial visualization of a project.

Overall, this body of work tries to stake a claim where these two attitudes find a common ground—and at times forcefully subvert this relationship. All through the monograph, drawings and pictures act as incubators for ideas, materials, and techniques which are generative. As often happens, the making of a book brought some clarity to a body of work done over a long span of time; these two volumes suggest that 'drawings' and 'pictures' are no longer distant poles of a spectrum within which architectural production might be placed, but are instead two realms which are moving closer and closer together.

简介

埃琳娜·曼费迪尼

"古典绘画大师已经意识到需保持所谓绘画整体的必要性，即绘画背后和三维空间展现的生动幻象之上的平面性的始终呈现。"

克莱门特·格林伯格[1]

《埃琳娜·曼费迪尼作品解析》收录了曼费迪尼事务所在过去6年的设计活动中创作的作品。这些作品没有按照时间排序，而是按照创作探索方式被分为两大类，收录在水平和竖直两册之中。这两个分册的标题暗示着两种不同的画面方向：横向和纵向。在艺术和画法几何中，人们将画面定义为位于眼睛和目标对象之间的一个抽象平面。在传统意义上，这个抽象平面垂直于人的视线。

当画面为纵向时，它的顶部通常对应着观察者的头部，而其底部对应观察者的足部。一个纵向的画面描绘一个场景空间，正如艺术家在直立时所看到的一样。画面的纵向不仅传达了创作者的视觉行为，同时也定义了观察者的观看方向。纵向画面限定了一种观察方式。而当画面为横向时，它相悖于观察者直立姿势的立体边界方向，变为了对于视觉体验的一种模拟。与竖向画面不同，横向的画面与其说是一种观察方式，不如说是一种置身于画中世界的途径。当纵向画面激发了以人类为主体的表述方式，横向画面则超越了观察者的眼界，投射于远大于人类尺度的场景之中。

曼费迪尼事务所的近期作品被划分为水平和竖直这两个主题，探讨了画面所能呈现出的两种截然不同的本质。这两个主题将成为彼此之间一个极富想象力的对照。它们不仅有关如何定位图像，还讲述了怎样生成图像以及使观察者参与其中。此外，画面从纵向到横向的旋转可以使图像的主体从抽象几何转变为具体形状。在这两个分册中，竖直呈现了我们惯例中的建筑图纸。这些图纸通过描述一组抽象规则（非图像）来构建几何框架，进而可以用来生成可观察的物体、效果或图像；水平则恰恰相反，其收录图像以可观察的物体为出发点，进而实现抽象的表述。

本书上册水平中收录的作品与人眼看到的水平方向的传统透视相比极为不同。这些横向图像是在正交投影下生成，它们无聚焦，无视角，无透视。它们与定义的状态或尺度无关，同时没有景深。这些作品的平面性将插图、动画、摄影、平面设计融合在一个艺术空间中，摧毁了它们的等级差异。他们的主题通常是照片般逼真的，每个描述的原始材料都是3D扫描采集的自然元素，如植物群和动物群。扫描产生点云和纹理，经过数字处理后被制作成绘画作品般的图片。

这些作品的表现形式是针对观众创作出的，并不是个人风格的实验。作品的成功取决于他们与广大公众进行戏剧性接触的程度。在这里，真实感与熟悉感不是缺点，反而被视为作品成功的关键因素。艺术和戏剧中出现的参与者（观众）的存在是舞台的根本问题，戏剧化使观众的融入成为可能。

在我看来，照片现实主义的表现形式与公众反馈之间的关联对建筑师而言非常有价值。这些图像表现出一种新的综合环境，在这种环境中，我们尤其关注物质的实体再现及其触觉效果。其中的熟悉感是现实中多次突变的结果，它成为作品与观众建立联系

---

[1] "Modernist Painting, Art and Literature", Spring 1965

时不可或缺的因素。因此，这些水平的画面作品造就了一种新的数字知觉的隐喻，其中真实和梦幻、熟悉和未知、静止和动态之间的界限变得越来越模糊。

水平这个主题传达了建筑表现中的照片写实主义的地位，为建筑学在现代艺术和数字媒介之间创造了新的图像空间。本书上册解决了图像能否囊括形象和叙事的问题。这些特殊的图片具备光学效果和分层的材料饰面。这些作品证明了平面具有前所未有的可能性，它可以营造感官效果，使人产生文化联想，并重新引发关于幻想在建筑中的作用的议题。就个人而言，这些作品是我们对已知的数字渲染的无声抗议。它们旨在证明：对于这个数字时代的建筑师而言，照片写实主义能够不仅限于商业用途，它还可以成为一种强大又令人印象深刻的创作工具。

本书下册竖直是建筑图纸的集合，主要是建筑立面图。这些作品探讨了复杂的编程线条在描绘建筑立面方面的潜力。该系列作品既是建筑研究也是独立艺术品。这些图纸可以理解为缩小建筑物的复制品，同时也可以理解为全尺寸印刷的艺术品。

编制一套数字草图的意义在于，在过去的20年中，计算机已经能够生成新的几何图形，这些几何图形不能再单独依靠诸如立面图、平面图和剖面图等传统的表现方式来理解。当今建筑的体量是多方位的，无法精确定向，因此，绘图惯例的规范变得含糊不清，甚至有些人觉得它们已经过时了。

竖直探索了如何改进图纸，尤其是建筑立面，以使其再次成为一种创造性工具而非还原工具。具体地说，这些作品中没有一个是从三维到二维转化得来的（而通常建筑图纸是从三维到二维的转化演变来的），它们是无限三维叠加的起点。这些图纸是灵活的，并且在抽象和自由的平面中以各种形态进行切割、粘贴、投影、挤压和折叠。这些图纸作品利用二维图形来蕴藏多种不同含义，同时创造了丰富的物理空间下的三维表现。它们反映了我们当代的数字可视化工具，以及它们的多维混合特性如何加速我们的创造过程。

这些立面图中的大部分是由代码逐张快速生成的，然后用这些立面图组成草图。这些立面图仍然有意地保留了抽象性，他们拥有墨水的色域，但更像是一张草图而不是真正的立面图。这些作品揭示了递归数字时代的规律，并展示了规律中的秩序感如何通过建造转移到各种尺度的设计当中。这些项目的特点是对几何重复系统的疯狂关注，并展示了一个将聚合逻辑、模式和网格与构造解决方案联系起来的领域。竖直通过对图纸引发思考的不断实践，提示人们绘制图纸这种教学方式不但不会过时，反而可以在当代被改进并用于激发我们的想象力。下册收录的作品传达了一种内在情感，它将心灵引导到一个尚未存在的现实。

从历史上看，建筑可视化一直是固有二分法的产物，因为它们通常被视为"艺术"能力与"技术"技能相结合的表达。它们把直觉和反思之间、具体和抽象之间的关系具体化。换句话说，它们在"想象一个可信的现实，或者简单作为一种达到目的的手段两种方式"之间切换。

随着数字革命的发展，图纸与图像的双重性质变得更加两极分化。我们既有技术图纸，也有逼真的渲染图，它们区分了可视化这一领域的专业人士和普通大众。随着数字时代的到来，图纸的图像内涵面临着一个共同的风险，即从过去特定创意的表达，转向项目的商业可视化。

总而言之，这一系列作品试图在这两种态度的共同关系中找到立足点，抑或强有力地颠覆这种关系。本书中图纸和图像代表了创造性思维、材料和技术的孵化器。正如经常发生的那样，一本书的制作是为了诠释很长一段时间内所完成的作品，本书表明，"图纸"和"图像"不再是放置建筑作品的光谱两极，而是两个越来越近的领域。

NOTE ON CODING AND GEOMETRICAL COMPOSITION

BY LI NING

The work of Atelier Manferdini relentlessly questions the audience's relationship with a given work. In this volume, the particular focus is on our orientation to the world in the form of a landscape. At their core, these landscapes interrogate the rhythm and the legibility of an underlying ordering system. This is, in some ways, a direct response to their subject: the natural world. There is a sense of order underpinning nature. One of the major contemporary ways to engage with this system is through carefully crafted coding. With the help of computer technology, complex algorithms can be implemented to suggest new geometrical possibilities for building envelopes, and thus realize the goal of creating new architectural relationships. Further, they can be used to describe entire compositions. The use of code and scripting can reveal the order inherent in nature, shifting the viewer's perspective and understanding of a scene. This exploration of compositional qualities and patterning possibilities is at the heart of the scripting used Landscapes.

To begin to understand the thought process behind the employ of digital algorithms, it is worthwhile to examine the historical reductionists' interpretation of nature. The reductionists believed that natural phenomena at a specific level of structure can be explained by analyzing the interaction of lower-level components. Acting in opposition to reductionism, holisticists believed that there is no violation of physical and chemical laws in nature, and that the whole is greater than the sum of the parts. When the parts are organically combined, new properties would be created. These new characteristics belong to the whole and cannot be derived from the separate analysis of each part. But this does not mean that reductionism has completely fallen by the wayside. For example, the latest progress on developmental biology—the phenomenon of life is the physical and chemical process controlled by gene participation and the control of genes is the result of the power of natural selection—gave holism a serious setback. The latest research results show that many of the previous behaviors and phenomena that seemed impossible to restore have indeed been reconstructed. Today's scientists are once again believers of reductionism.

However, this does not mean that holism has been completely abandoned. Scientists believe in reductionism's tenet that phenomena can be examined through their constituent parts, but they have a holistic point of view that understands the whole encompasses characteristics which cannot be understood solely by looking at its parts. Modern research on nature is not an isolated part of a single study, but a more macroscopic study of the connections and interactions of various parts. The natural processes resulting in life are extremely complex, reversible physical and chemical phenomena. These processes point to a fundamental sense of order which underlies not only science, but many aesthetic disciplines as well. These reversible rules can be written into algorithms to generate architectural forms and create new kinds of compositions by understanding the importance of both the whole process, as well as its components.

We can see the essence of the inherent order of nature by describing its rules in the form of algorithms. These algorithms are collections of instructions and methods for composition. As a collection of instructions, they relate to both the standard analog design flow and the digital design flow. But in the field of digital design, algorithm design has a special meaning. By directly writing and modifying the code rather than the form, the designer uses the programming script language to enable the possibility to go beyond the limitations of the software user interface. Generally, computer programming languages used in algorithm design include Python, C#, VB, MEL (Maya Embedded Language), 3Dmax Script (3Dmax Embedded Language),

Rhino Script (Rhinoceros Embedded Language), Java, and the like. Algorithm design fully exploits the computer's ability as a search engine and performs some unusually time-consuming tasks. As a result, it allows for certain tasks that go beyond the standard design constraints to be completed.

The development of computer technology has broadened our way of thinking about the implementation of digital design. Whether it is through an existing algorithm, a rule system, or a new script discovered through research, previously impossible designs can be realized with the help of computational technology. In particular, models that require a large amount of computation can be quickly established. Digital technology helps achieve many forms that were previously difficult to express and simulate geometrically. Furthermore, it has the potential to anticipate future possibilities of architectural form through models, so that the simulation has diachronic characteristics which allow it to be easier to manipulate.

Professor Xu Weiguo from the School of Architecture at Tsinghua University argues that one of the major principles when working digitally is to design variables. Each field of input is an analysis of one or more important influencing factors in the design process. The result is changed by altering the value of that input variable, or by changing the relationship between the inputs. This forms the basis for a design methodology which engages with scripts to create specific output relationships. Thinking about design in this way is a continuation of the interplay between reductionism and holism: there is a consideration of both the discrete input variables and the script on the whole.

The ideal design practice for creating scripted architecture is to first investigate the base and to collect the initial data. The second step is to establish an algorithmic model, transform the data from the previous step into input variables, analyze the variables, and extract the architectural design. The third step is to use the preliminary scripted model to continuously modify, adding more variables to make the model's logic more relevant and the generated form more diverse. The fourth step is to use this scripted logic compositionally, completing the aesthetic intent by applying a prevailing sense of order. Each variable that is added, and each connection that is made between them, is an opportunity for applying another layer of information—and of meaning.

The aesthetic disciplines share the same base desire as the scientific discipline to find and understand the organizational principles between forms. In the ideal case, the relationship between these forms and laws is mathematically described. Mathematical methods for describing natural forms have been in full swing, but the degree of simulation is not equal to the visual characteristics. Taking painting as an example, the analogy is: the lower the degree of abstraction, the stronger the image resolution, the more features that can be displayed. But at the same time, it is less likely to distill and spotlight some of the most essential underlying structures. On the contrary, the higher the degree of abstraction, the easier it is to highlight the most essential structural features and the stronger its symbolic function. But if the painting is too abstract, it is difficult to form clear connections, possibly causing confusion and ambiguity. Balancing both fidelity to the visual characteristics of what is being represented and a certain degree of abstraction, allows for the work of art to best render its subject and the structural principles behind it.

The same is true for the building shape generation based on scripted codes. It is necessary to adapt to the environment and characterize its internal laws. The resulting form must have typical visual features. The process of building a structure follows the laws of form that the scripted codes suggest, but the results are not necessarily "like" creatures. In this space between what would be expected based on the algorithm and the expressed architectural representation is the moment for aesthetic composition. The base order the script offers becomes both a launching off point, and something to continually reference.

Algorithms, scripting, and digital technology provide an underlying system for architectural designers which has both newfound aesthetic possibilities and a recognizable sense of order. There is a certain amount of calibration that is required to balance visual fidelity and abstraction, but the result is a rich blend which offers a large amount of potential for composition. Controlling both the input variables, and the relationship between them, allows for an endlessly iterative process with which to design. This plays out in Elena Manferdini's work inscribed herein. Each landscape is a new viewpoint on nature, one that questions how we might understand order and how we, as the audience, may position ourselves in relation to the composition.

浅论编码和几何构图

李宁

曼费迪尼事务所的工作意在挑战读者对特定作品的认知，在本书中，其特别关注的是世界观。本书的核心是讨论潜在规则系统的节奏和易读性，在某些方面，作品直接回应自然世界。

本书作者相信有一种秩序感支撑着自然，而参与此秩序的主要方式之一是通过精心设计的编码。在计算机技术的帮助下，可以通过复杂的算法来构建包络的新几何，从而实现创建新建筑关系的目标。此外，算法可用于描述整个组合物，代码和脚本的使用可以揭示自然界固有的顺序，改变观众的视角和对世界的理解。对构图和图案化多种可能性的探索是本书的核心。

为了理解算法应用背后的思考过程，有必要研究历史还原论者对自然的解释。还原论者认为：通过分析较低层次成分的相互作用，可以解释特定结构层面的自然现象。整体主义者则反对还原主义，认为自然界没有违反物理和化学法则，但是整体大于各部分的总和，当部分有机组合时，将拥有新属性，这些新特征属于整体，不能从每个部分的单独分析中得出。但这并不意味着还原论已完全被击垮。例如，发育生物学的最新进展——生命现象是由基因参与控制的物理和化学过程，基因的控制是自然选择力量的结果——给整体论带来了严重的打击。最新的研究结果表明，许多以前似乎无法还原的行为和现象确实得到了重建。今天的科学家们再次相信了还原论。

然而，这并不意味着整体主义已被彻底抛弃。科学家们相信还原论的初衷是可以通过其组成部分来检验现象，但是他们有一个整体的观点，即接受只有整体包含的特征，这些特征只能通过研究其各个部分来理解。关于自然的研究不是单一地研究孤立部分，而是对各部分的联系和相互作用的宏观研究。产生生命的自然过程极其复杂，但是却是可逆的物理和化学现象。这些过程指向一种基本的秩序感，它不仅是科学的基础，也是许多美学学科的基础。这些可逆规则是可以写入算法中的，可以通过了解整个过程及其组件的重要性来生成建筑形式并创建新类型的组合。

通过以算法的形式描述其规则，我们可以看到自然固有秩序的本质。这些算法是合成的指令和方法的集合。作为指令集，它们与标准模拟设计流程和数字设计流程相关。但在数字设计领域，算法设计具有特殊的意义。设计人员通过直接编写和修改代码而不是菜单来进行创作，使其有可能超越软件用户界面的限制。通常，算法设计中使用的计算机编程语言包括Python、C#、VB、MEL（Maya嵌入式语言）、3Dmax脚本（3Dmax嵌入式语言）、Rhino脚本（Rhinoceros嵌入式语言）、Java等。算法设计充分利用计算机作为搜索引擎的能力，并用其执行一些异常耗时的任务。因此，它能够完成超出标准设计约束的某些任务。

数字技术的发展为算法设计的实现扩宽了思路。无论是已有的算法或者规则系统，还是通过研究而发现的新的形态生成算法，都可以在数字技术的帮助下得以实现。尤其是需要大量计算的模型，通过数字技术的帮助可以快速建立，数字技术帮助人们实现了很多以前在几何学上难以表达和模拟的形态。不仅如此，数字技术还可以通过模型来预测建筑形态的未来，使建筑形态的模拟具有历时性的特点，增加了人们对于形态生成的可控制性。

清华大学建筑学院徐卫国教授认为：参数化设计是把设计参变量化，每个参变量都是对设计过程中的一种或多种重要的影响因素的解析，改变参变量的数值会带来设计结果的改变。每个参变量自身与不同的参变量之间是两个不同的影响设计结果的因素：一个是参变量本身数值的变化，另一个是参变量之间的构成关系。参变量数值和参变量之间的构成关系的变化都会改变最终结果。参数化融入建筑设计便形成了参数化设计，参数化设计的方法是设计结果受到参变量数值以及它们之间构成关系两方面控制的设计方法。

参数化建筑设计的理想设计方法第一步是对基地进行调研和对基础资料进行搜集；第二步是建立参数化模型，将上一步的资料转化为参数，对参数进行分析，提取对建筑设计起到关键作用的重要参数，将其输入计算机形成模型；第三步是利用初步的参数化模型生成"基本建筑形体"，并在此基础上不断对参数化模型进行修正和补充，加入更多的参数，使模型的生形逻辑更加贴切，生成的形式更加多样化；第四步是按照设计要求对建筑形体进行回馈和检验，以满足其原始的要求。

美学学科与科学学科具有相同的基本愿望——发现和理解形式之间的组织原则。在理想情况下，这些形式和规律之间的关系可以用数学方法描述，但是其模拟仿真的程度与视觉特性并不是对等的。以绘画为例做类比：抽象程度越低则意象越强，其绘画性特点越强，就能显示越多的特征，但与此同时也越不容易突出或者掩盖、歪曲某些最本质的结构特征，符号性的象征功能就越差；相反，抽象程度越高的意象越容易突出事物最本质的结构特征，其符号象征功能就越强，但由于这种意象太过抽象，其绘画性特点就较弱，与具体事物很难形成明确的联系，容易引起模糊与模棱两可的认知。在艺术中，理想的意象是与希望要表现的概念同构的、模拟仿真程度不高的意象，这种意象既要具有较典型的事物的视觉特征，又要有一定程度的抽象性。基于脚本代码的建筑物形状生成也是如此，既要适应环境、表征其内部规律，由此生成的形态又要具有典型的视觉特征。

算法、脚本和数字技术为建筑设计师提供了一个基础工具，它具有发现新美学的可能性，也可以发现新秩序感的可识别性。平衡视觉保真度和抽象需要一定量的校准，但结果是丰富的、多样性的，也为设计师提供了大量的组合潜力。控制输入变量以及它们之间的关系可以使设计无休止的迭代下去。

上述观点在埃琳娜·曼费迪尼的作品中有所体现，每个作品都是一种新的自然观，一种质疑我们如何理解秩序的观点、一种作为观众如何理解自己与作品关系的问题。

LANE CRAWFORD
SPRING SUMMER 2018

Date: 2018

Client: Lane Crawford

Location: Beijing/Hong Kong/Chengdu/Shanghai, China

连卡佛
2018年春夏

时间：2018年

客户：连卡佛

地点：中国，北京 / 香港 / 成都 / 上海

Lane Crawford is a retail company which sells luxury goods founded in Hong Kong in 1850, Lane Crawford has more than 50,000 square meters of retail space, including several physical locations throughout China, in addition to an online store. Lane Crawford strives to discover new forms of expression through their global exclusive collections, unique art collaborations, and partnerships with emerging local talent. As part of their global program "What is Art for You", in 2018 the company Art Director asked Elena Manferdini to reimagine the shopping windows, pop-up stores, and entry lobbies of their 8 stores in China. Atelier Manferdini designed 8 digital landscapes which come alive through custom Augmented Reality Applications. While immersive vinyl printed landscapes act as a theatrical backdrop for the clothing, a series of flat screens run the digital animations of fields of flowers and insects moving in the wind. Viewers from outside the shopping window can look at the clothing through their smart phones and can use the app to see the insects flutter around the mannequins. The viewer can also take pictures, record videos and share their augmented reality experience online.

连卡佛是一家奢侈品零售公司，于1850年在香港创立，目前拥有超过50000平方米的零售空间，包括各地的几家实体店和一家线上商店。连卡佛致力于通过其全球独家收藏、独特的艺术合作形式以及与本地新秀设计师的联手，探索新的表现形式。2018年该公司艺术总监邀请埃琳娜·曼费迪尼参与其全球项目"什么是艺术"，要求设计师重新设计其位于中国的8家商店的橱窗、快闪店和入口大堂。最终，埃琳娜的工作室完成了8个数字景观的设计，通过AR（增强现实）应用程序在手机上使其栩栩如生。店内服装以喷涂了绿色乙烯基材料的人造地貌为背景，一系列屏幕上展示的是舞动的花海和蝴蝶。消费者在橱窗外可通过智能手机观看服装样式，还能够在同款应用程序中看到昆虫在模特周围飞舞。用户可以拍照片、视频发布到网络上，分享他们的体验。

Blueberries 蓝莓

Daisy 雏菊

Although architects may come to terms with the role of subjectivity and its importance and relevance for any creative process, they are often culturally highly biased against beauty and the aspiration for generating architecture that will be focusing on intentionally generating the experience of beauty. Elena Manferdini does not share this view. Projects such as 'Lane Crawford' of 2018, in describing nature, evoke human emotions—as does nature, though she depicts it electronically and builds up the drawing synthetically—becomes part of the magic through the architectural spaces that she designs. The use and evocation of poetic images in the work turns the design into an emotive one. The use of the iPhone App she developed to bring in the singing birds and the flying butterflies makes the experience more personal as the observer interacts with the animated imagery.

Yael Reisner

尽管建筑师可能意识到主体性的作用，以及它对任何创作过程的重要性和相关性，但他们往往在文化上对美有高度偏见，并渴望创造出注重有意识地创造美的体验的建筑。埃琳娜·曼费迪尼并不同意这种观点。以2018年连卡佛"什么是艺术"项目为例，她用电子方式描述自然，并合成了图画，她设计的建筑空间成了魔法的一部分。诗情画意的图像使得该设计极为感性。观众使用她开发的手机应用程序，可以看到歌唱的小鸟和飞舞的蝴蝶，使观察者与动画图像交互时的体验更加个性化。

雅尔·雷斯纳

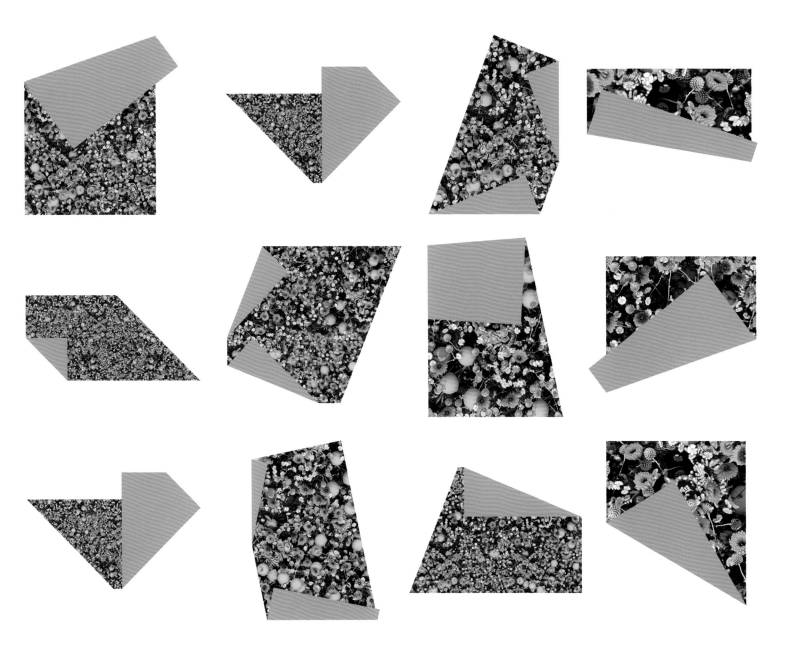

Folds 折叠

21

Delphinium 飞燕

Folds 折叠

23

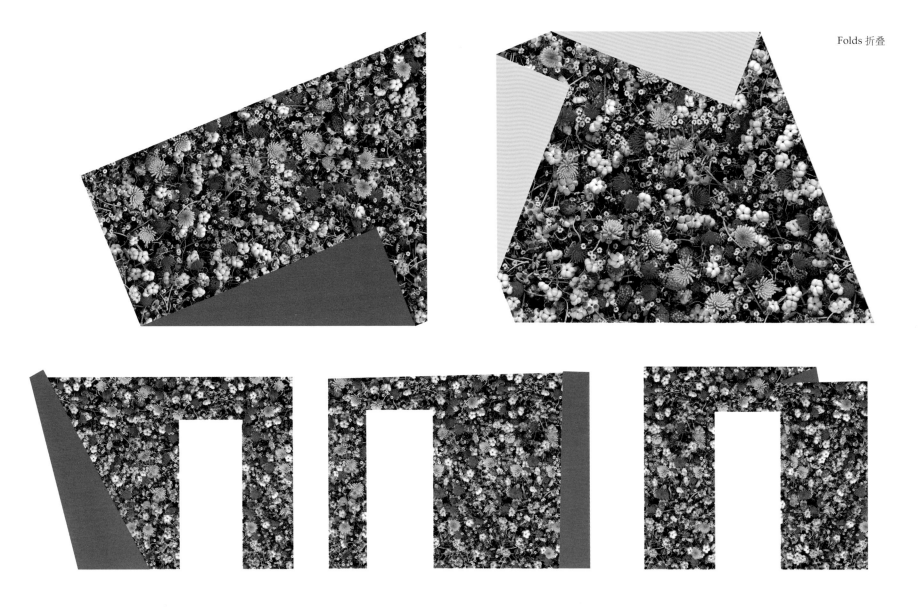

Folds 折叠

Folds 折叠

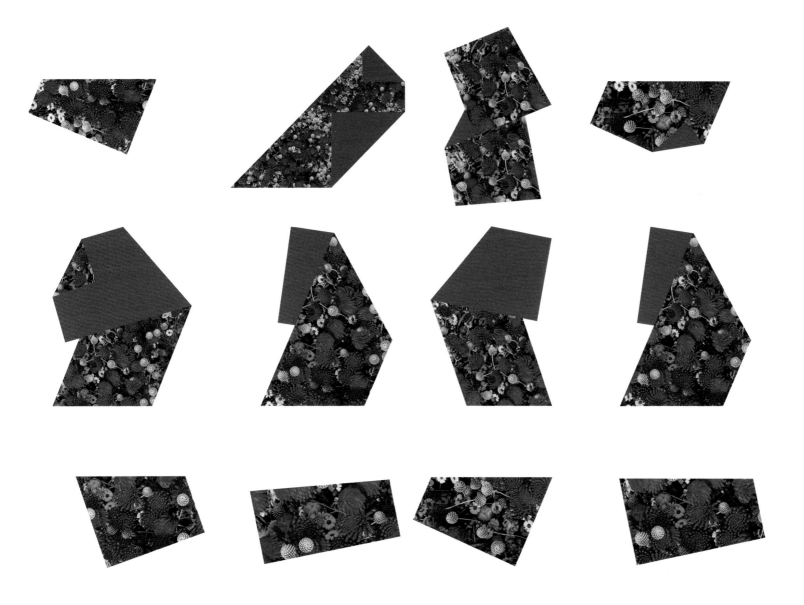

27

Folds 折叠

29

Shanghai Window 上海橱窗

Shanghai Window 上海橱窗

Shanghai Window 上海橱窗

Digital transformation does not simply require new technology; it requires a new, more agile mindset. Every line of business must have access to the digital tools needed to innovate at the edge. The job of the architect, in this case, is to provide and invent new applications.

Augmented reality adds virtual elements to an existing reality, rather than creating a reality from nothing. Architecture will be viewed with added information and enhance our aesthetic experience. It can also be a pedagogical experience. Elena Manferdini strives for integration between people and architecture, rather than isolation. Virtual reality isolates the individual in an artificial world, whereas augmented reality allows you to interact with the real world directly. The use of an AR device could be projected through multiple mediums such as lenses or directly through phones.

Gabriel Esquivel

Beijing Window 北京橱窗

数字化转型需要的不仅是新技术，还需设计师拥有新颖的、更加敏捷的思维方式。各个行业从业者都必须掌握在该领域创新所需的数字化工具。就此意义而言，建筑师需要做的便是提供和发明新的应用程序。

增强现实（AR）技术将虚拟元素引入现实，而非从无到有创造现实，它赋予建筑以新的生命力，增强了人们的审美体验。它甚至也是一种教学体验。埃琳娜·曼费迪尼致力于达到人与建筑的和谐统一，而非疏离。虚拟现实（VR）技术将人们困在人造世界中，AR技术却实现了人与现实世界的互动，而这一过程仅需借助如相机或手机这样的媒介就能实现。

加布里埃尔·埃斯奎维尔

URBAN FABRIC RUGS
BUILDING PORTRAITS RUGS

Date: 2018

Client: Urban Fabric Rug

Location: Shanghai, China

"Urban Fabric Rugs" creates abstracted depictions of our built environments by taking flat urban street grids and conveying them in extruded reliefs with soft hand-knotted and carved wool and silk. This collection of textiles was created in collaboration with Urban Fabric Rugs, and is an adaptation of Manferdini's drawings and images from her body of work entitled "Building Portraits".

The compositions celebrate uniqueness and convey a sense of place. Each rug questions and repositions our understanding of the make-up of the urban environment. One piece, "Coal on Smog", utlizes grey-on-grey tones to comment on the smoggy air quality of the Shanghai. The layering of colors and geometry imply the various social, cultural, and spatial complexities of the cities on which they are based. These pieces have entered the permanent collection of LACMA.

城市肌理地毯
建筑肖像地毯

时间：2018 年

客户：城市肌理地毯

地点：中国，上海

"城市肌理地毯"系列作品用手工编织羊毛和丝绸，用浮雕的形式表现平面城市街区，抽象地展现了我们的建筑环境。该系列作品与"城市肌理地毯"的设计师品牌联名创作，是由埃琳娜·曼费迪尼的系列作品"垂直建筑"中的绘图改编而来。

这些作品颂扬独特性，展现出浓厚的地域风格。每块地毯都会使人们思考和重新定位我们对城市环境构成的理解。其中一件名为"雾霾中的煤炭"的作品，使用了不同色调的灰色表现上海烟雾弥漫的空气状态。富有层次的色彩和几何图案意味着上海这座城市的社会、文化和空间复杂性。该系列作品已被洛杉矶博物馆永久收藏。

Building Portraits Rugs　建筑肖像地毯

When architects deployed CAD-CAM systems, turning the technology into a manufacturing reality, they became the craftsmen, since they skillfully and digitally create the drawings that turn directly to physical objects. Greg Lynn, with his Alessi Tea and Coffee Piazza, 2000, was perhaps the first architect to deploy this technology for design.

Elena Manferdini, who finished her architectural studies in that year, was one of the pioneers in channeling this craft towards fabricating textiles and dresses for the leading fashion houses' haute couture shows; an architect activity that continues today and has become a tradition since architects seem to be the best at digital drawing.

Yael Reisner

当建筑师们开发了CAD-CAM系统，并将这项技术应用于实际时，他们成了工匠，因为他们能够熟练地、数字化地创作出可以直接转化为实物的图纸。格雷格·林恩于2000年设计了他的阿莱西茶和咖啡广场，他可能是第一个将这种技术应用于设计的建筑师。

也是在2000年，埃琳娜·曼费迪尼完成建筑专业学业，率先使用CAD-CAM系统为顶级时尚品牌高级定制时装秀设计纺织品和时装。时至今日，建筑师参与时尚设计已成为一项传统并延续至今，因为建筑师们似乎骨子里就擅长数字绘画。

雅尔·雷斯纳

Building Portraits Rugs 建筑肖像地毯

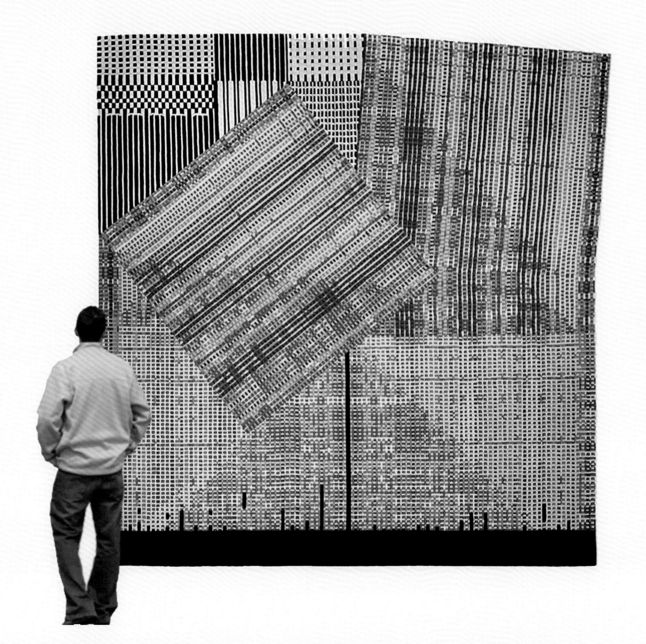

Silk Textiles 丝绸纺织品

FRAMED SKYLINES

METRO

Date: 2016

Client: LA Metro

Location: Los Angeles, California, USA

天际轮廓线

地铁

时间：2016 年

客户：洛杉矶地铁

地点：美国，加利福尼亚州，洛杉矶

"Framed Skylines" is one of the 4 final proposals selected by Metro for the facade of their new facility building on 6th and Alameda in the Arts District in Los Angeles. Metro defines how a city is perceived. As these transit lines carry passengers through the built environment, they provide distinct windows onto Los Angeles. The role of Atelier Manferdini's artwork is to add to the narrative that there is something more to the Metro than simply moving people around: it shapes people's daily lives and experiences of the city.

The alternating pattern is a reinterpretation of the pattern created by railway tracks. The matte black panels create a direct reference to the railway, while the reflective pieces in between begin a conversation with the context. As someone moves down Santa Fe Avenue in front of the Metro Building, they will feel the rhythm and motion of the train transposed onto the facade. Each reflective panel is framed between vertical black strips, highlighting a series of specific views of the city. This ultimately creates an experience much like the Metro itself: a collection of ever-changing glimpses of Los Angeles.

在洛杉矶艺术区第六大道的地铁设施楼外观设计中，"天际轮廓线"从最终的4个设计中脱颖而出。地铁定义了人们感知城市的方式。不同的地铁线路在带领乘客穿梭于城市中的同时，也给人们提供了欣赏洛杉矶的视角。此次设计的目的在于增加地铁的叙事性，证明地铁不仅能运载乘客，也深入人们的日常生活，影响人们对城市的感知。

交替模式是对铁路轨道形成模式的重新解读。两排哑光黑色面板使人联想到铁轨，中间反光板的加入使得整个设计与背景形成对话。当人们行走在大楼前面的圣达菲大道时，便会感受到列车穿梭时的节奏和韵律。镶嵌在垂直的黑色面板之间的反光板，倒映着这座城市的一系列特定风景。这一设计最终创造了一种在地铁上体验洛杉矶的经历，即人们可以从中欣赏到千变万化的景观。

Framed Skylines 天际轮廓线

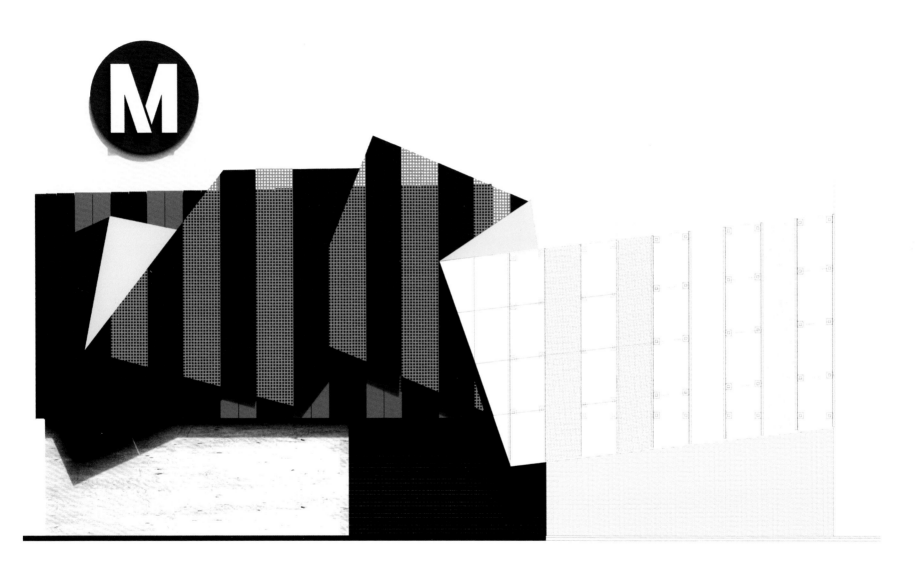

Elevation 正面

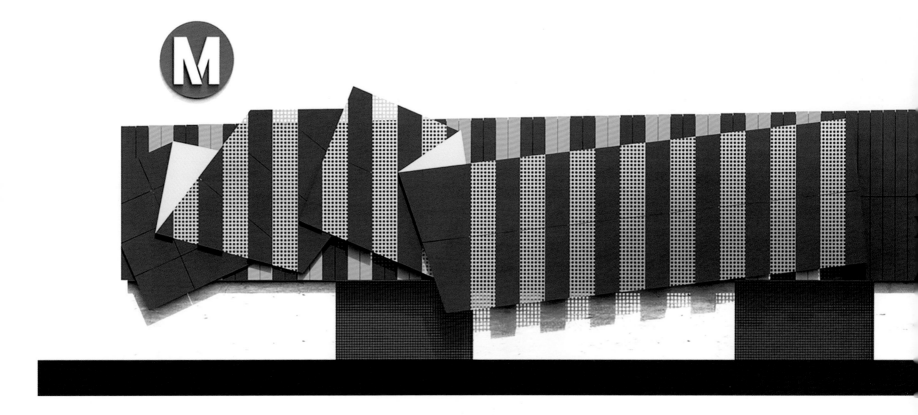

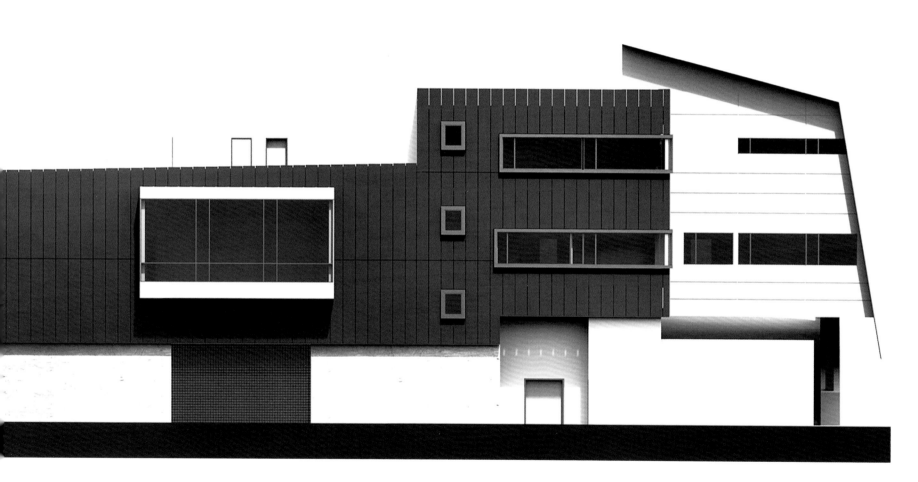

Elena Manferdini's work is strictly and almost obsessively architectural. It's not pictorial in a general sense and it's not art. Its only way to be correctly understood is to be appreciated in the tradition of architecture. If addressed as graphics or painting, it falls short. However, if addressed as architecture, it not only rises to the occasion of contemporary architecture but it also rises to the occasion of architecture with a political projection. Her work has an interest in art and an interest in color. However, this belongs to an architectural tradition that derives from her knowledge of Italian architecture, engineering, assembling a building, and then working them out two-dimensionally through drawings.

Jeffrey Kipnis

埃琳娜·曼费迪尼的作品十分痴迷于建筑性。它并非一般意义上的图画，也不是艺术。正确理解其建筑设计的唯一方法就是欣赏建筑的传统。如果作为图形或绘画处理，则是失败的。然而，如果称之为建筑，那它不仅达到了当代建筑层面，而且还上升到了具有政治映射的建筑层面。她的作品表现出对艺术和色彩浓厚的兴趣，这种风格源自她的建筑传统知识：对意大利建筑、工程、组装建筑的掌握，并通过绘图将这些设计跃然纸上。

杰弗雷·基普尼斯

Diagrams 分析图

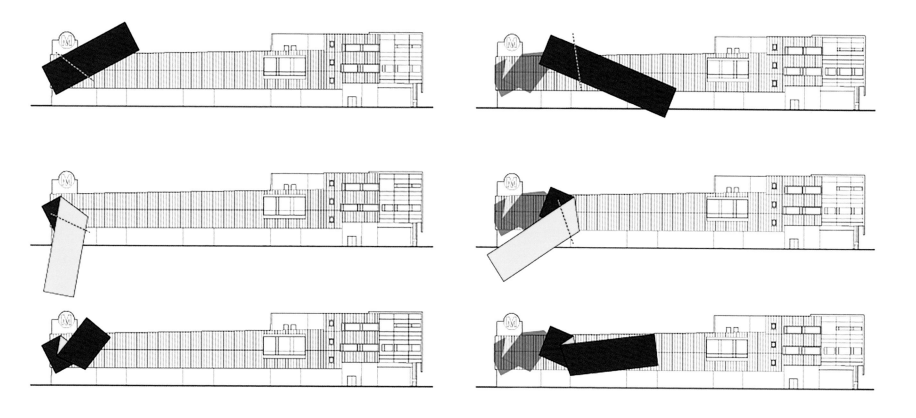

LOO WITH A VIEW
RESTROOMS FOR THE LONGMONT MUSEUM

Date: 2015

Client: Longmont Museum

Location: Longmont, Colorado, USA

观景洗手间
朗蒙特博物馆洗手间

时间：2015 年

客户：朗蒙特博物馆

地点：美国，科罗拉多州，朗蒙特

The Art in Public Places Commission of the City of Longmont, selected Atelier Manferdini to design and enhance the restrooms in the new auditorium of the Longmont Museum with permanent artwork.

"Loo with a View" takes the restroom of the Longmont Museum into another realm with a landscape that blurs the realistic and the fantastic. The concept is derived from Ancient Romans who painted their villas with natural lanscapes which celebrated poets, artists and writers. The *Locus Amoenus*—which is the Latin expression for an imaginary place of meditation—was described by Lucrezio in his *De Rerum Naturae* as the space where one can escape from the hectic grind of metropolitan life. Therefore, when landscapes or "*Loci Amoeni*" are represented in villas, they embody the ideal meaning of a space as a place of meditation where one could find his/her own thoughts.

When going to a restroom, for some, it often is their moment to escape. The imagery on the wall builds on this Roman villa tradition in order to aid in the creation of a pervading sense of calmness and peace.

位于美国科罗拉多州朗蒙特市的公共场所艺术委员会邀请了埃琳娜·曼费迪尼的工作室以永久性的艺术品设计和改善博物馆新展厅的洗手间。

该设计名为"观景洗手间"，它将朗蒙特博物馆的洗手间带入了另一个如梦似幻的境界。这一概念借鉴了古罗马人的做法，他们喜欢用自然景观装饰住宅，这些景观多歌颂诗人、艺术家和作家。卢克雷齐奥在他的作品《自然之路》中曾这样定义"安乐之所"：在这里，人们可以忘记繁忙的都市生活。"Locus Amoenus"在拉丁语中意为"一个想象中的冥想场所"。因此，当这一概念被引入住宅设计后，它们赋予建筑新的意义，即理想的冥想场所的意义。置身其中，人们可以找回自我。

洗手间里的片刻时间，对于一些人来说，往往是他们逃离尘世的时刻。洗手间墙上的图像以罗马住宅装饰为灵感，旨在为使用者营造出安静平和的氛围。

Digital Print 数字印刷

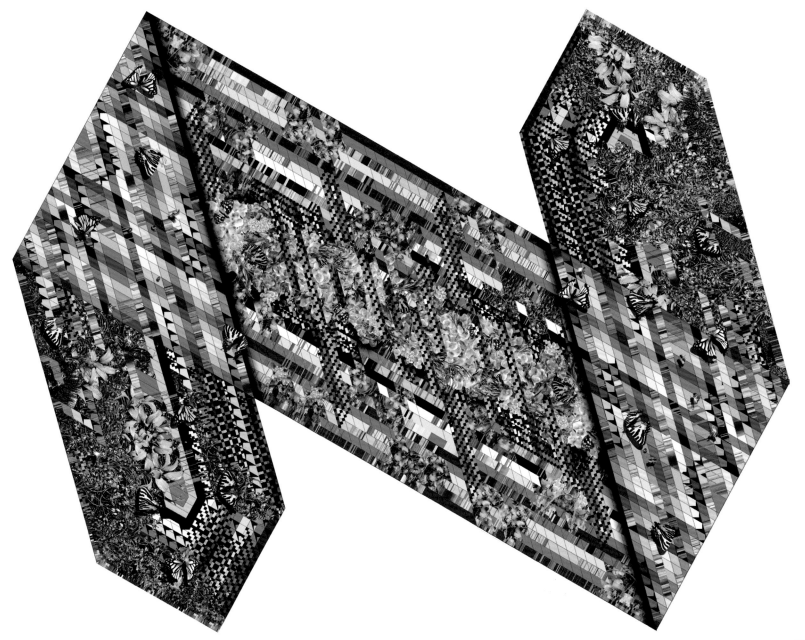

Elevation and Reflected Ceiling 正面和天花板

Color is one of the most influential phenomena in people's lives. That is why color can stimulate the mind in terms of creating joy or sadness. Although the perception of color is an individual and subjective process, cultural factors also influence how color affects us. Elena's use of color comes at the end of a post-affect, post-sensation moment. The early stages of the digital, the first decade of the 21st century when architecture became interested in discussing affect and sensation, color was part of that discussion. Her use of color comes from a current aesthetic concern, her interest in defining what is our cultural understanding of color in the 21st century, its relationship to Pop-Art, and other movements.

Gabriel Esquivel

色彩是人们生活中最具影响力的现象之一。那是因为颜色可以刺激大脑产生快乐或悲伤的感觉。虽然对颜色的感知是个体主观的过程，但文化因素也会影响颜色对我们的影响。埃琳娜对颜色的使用始于2010年。数字时代的早期，即21世纪的前10年，建筑学界开始热衷于讨论人的情感和感觉，颜色也成为这些讨论中的一部分。她对于色彩的运用来自当时的美学趋势，她对定义21世纪不同文化对于不同颜色的理解，以及它与流行艺术和其他的艺术的关系非常感兴趣。

加布里埃尔·埃斯奎维尔

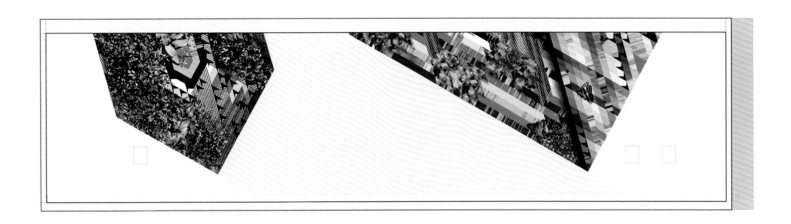
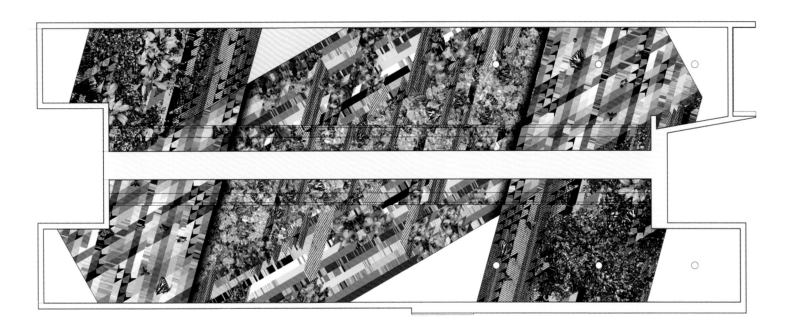

Installed Piece 安装件

MASSIVE PROJECTIONS
5TH GWANJIU DESIGN BIENNIAL

Date: 2013

Curator: Christina Kang

Location: Gwanjiu, South Korea

巨幕投影
第五届韩国光州设计双年展

时间：2013 年

策展人：克里斯蒂娜·康

地点：韩国，光州

"Massive Projections" is a site specific installation for the 5th Design Biennale in Gwanjiu, Korea. The wallpaper graphic design addresses issues of virtual perception, the framing of nature, and architectural space through the creation of an architectural bi-dimensional canvas. The subject of the printed images reinterprets the topic of Nature through a contemporary use of digital scanning techniques and scripting algorithms.

The graphic of the wall, although super-flat, is able to produce intense three dimensionality because of the estranged vantage points (a mix of orthographic elevations and plan views) of the 10 masses with their different natural paintings. These 10 volumes are "presences" rather than "sculptures" and they create a large event space through the simplicity of their compilation and their large scale.

The aesthetic sensibility of the project is generated from the existing fabric of the historical buildings and streets that acted as the source for the scripted geometry and chromatic palette of this new infrastructure.

"巨幕投影"是第五届韩国光州设计双年展的特定场地装置。墙纸平面设计通过创建一个建筑二维画布，解决了虚拟认知、自然框架和建筑空间的问题。通过数字扫描技术和脚本算法的使用，印刷图像重新诠释了自然这一主题。

墙面设计虽然是标准的平面，但由于这10卷自然主题绘画有着不同的前后位置关系（混合着立体和平面图形），依然能够产生强烈的三维感。这10卷是"平面"而不是"竖直"，设计师们通过简化它们的编绘和规模，创造出了一个大型的活动空间。

该项目的美感来自历史建筑和街道的现有肌理，这些结构是一些新基础设施的图案样式以及色彩的来源。

Wallpaper 壁画

In Representation, the focus is Figuration and Abstraction. Elena Manferdini's work is quite complex and high-risk in this regard. Traditionally, figuration is quite different from abstraction, since one is based on the representation of reality, while the other is based on 'essence.' Elena Manferdini brings together both elements within a figurative-abstract paradox. The abstract part of her work becomes an autonomous reality. Figuration in her work is the connection to certain recognizable aspects; it becomes the link to her artistic references. Consequently, the differences between abstract art and figuration are appreciated more and become more relative—as well as apparent and ill-defined. To speak of an abstract expression does not mean to invalidate or fail to implicitly recognize a certain degree of figuration. Inversely, an approach based on something real does not mean that it does not involve a certain abstraction.

Gabriel Esquivel

在表现手法中，具体和抽象是重点。埃琳娜·曼费迪尼在这方面的工作是非常复杂的，风险也很高。从传统上讲，具体与抽象完全不同，因为前者是基于现实的表现，而后者则基于"本质"。埃琳娜将两个相悖的元素汇集到了一起。她作品中抽象的部分成了一种自主的现实。她作品中的形象是和某些特定的可识别的元素相连接的，这成为她艺术参考的来源。抽象艺术和造型之间的差异被越来越多的人所欣赏，变得更加相关、更加明显和不明确。谈起一个抽象的表达并不意味着不承认某种程度的具体。同样，基于真实事物的方法并不意味着它不涉及某种抽象。

加布里埃尔·埃斯奎维尔

Wallpaper 壁画

FRESCO
DESIGN MATTERS GALLERY

Date: 2012
Client: Design Matters Gallery
Location: Los Angeles, California, USA

壁画
设计画廊

时间：2012 年
客户：设计画廊
地点：美国，加利福尼亚州，洛杉矶

"Fresco" is a site specific installation for the Design Matters Gallery entryway. The topic of the installation updates the classical exercise of "Still Nature" and uses it as a testing ground for a contemporary representation of matter. In particular, "Fresco" investigates how paint behaves when morphing from a photorealistic depiction of nature to a disfigured one. When colors mix, the geometrical matter that constitutes the photorealistic flowers and their details disappear into a primitive paint clog that can be mixed and brushed again and again into new colors and shapes.

The entryway of the gallery acts as a threshold from a monochromatic Los Angeles urban landscape into a multicolored immersive synthetic garden. The inside surfaces of Design Matters are covered by printed adhesive vinyl, where plants, flowers and insects occupy floors and walls. The interest of the installation is an expanded, hybrid nature whose depictions collapse reality and artifice, and insinuate that contemporary materials are often a mutation from the "original", producing a world in which fact, fiction, and fantasy co-exist.

"壁画"是专为安装于画廊入口处而特殊设计的。装置的主题重现了"静止的自然"的经典作品，并将其用作当代物质表现的试验场。"壁画"这一作品特别研究了绘画从对自然的写实描述演变为非写实描绘的过程。当颜色混合时，构成真实感花朵的几何物质和它们的细节就会消失在原始的颜料块中，这些颜料块可以一次又一次地混合和改变成新的颜色和形态。

画廊的入口作为展览的开始之处，引导人们从单调的洛杉矶城市景观进入一个五彩缤纷的沉浸式合成花园。装置材料的内部表面被印刷上去的乙烯基胶覆盖，植物、花卉和昆虫占据了地板和墙壁。装置作品的亮点在于地板和墙壁上满是植物、花卉和昆虫。该设计的特点在于它是一个扩展的、混合性质的自然，它的设计打破了现实和技巧，并暗示了当代材料往往是"原始"的变异，产生了一个现实、虚构和幻想并存的世界。

Close Up Printed Vinyl 特写玻璃纸

Digital Print 数字印刷

There must be a much tighter affinity between Nature and Culture within the built environment that people live in and are surrounded by. Remembering that humans are an equal part of the equation, we need to make sure that we design a humanized architecture for them.

It is evident that the loss of the aspiration to create the experience of beauty in architecture has, since the 1940s, often resulted in an environment that is alienating; not focusing on what makes architecture pleasurable.

Yael Reisner

在人们生活和建造的环境中,自然与文化之间肯定有更紧密的联系。要记住,人类是平衡中很重要的一部分,我们需要确保设计出一个非常人性化的建筑。

很显然,自20世纪40年代以来,人们失去了在建筑中创造、体验美的诉求,这就造成了一种疏远的环境,人们也开始不在意究竟什么能让建筑更加赏心悦目。

雅尔·雷斯纳

FABULATIONS
ISTANBUL DESIGN BIENNIAL

Date: 2014

Client: Zoë Ryan

Location: Galata Greek Primary School, Istanbul, Turkey

虚构情节
伊斯坦布尔设计双年展

时间：2014 年

客户：佐伊·莱恩

地点：土耳其，伊斯坦布尔，加拉塔希腊小学

"Fabulations" is an installation which explores the creative possibilities and some of the cultural implications of photorealistic representations obtained from high-definition reality-based 3D scans, in combination with the most advanced digital tools applied at the scale of architecture. With the evolution of contemporary technologies, and the consequent ability to interact with analog matter in digital environments, faithful depictions of nature can now be used to investigate new experiences that embrace realism, familiarity, narrative, and involvement of the audience as crucial ingredients. This technology re-opens the debate on representation by allowing a new kind of digital photorealism: the living picture.

The work embodies a new generation of synthetic environments, in which special attention is paid to the literal reproduction of matter and its tactile effects, where familiarity is the result of multiple mutations from reality and becomes an indispensable ingredient to establish a connection with the audience.

"虚构情节"是一个装置，它通过精准的高清3D扫描，结合应用于建筑领域的先进数字化工具，得到逼真的影像，进而探索这些影像的创新可能性以及文化内涵。随着现代技术的演变以及数字场景与模拟事物互动的必然性，可靠的自然描述现如今能够被用于研究更贴近现实主义、认知、叙事性和观众参与等关键要素的新体验。这项技术通过一种新型的数字照片写实主义——生动的画面，重新开启了关于表现主义的辩论。

这部作品体现了新一代的人造合成环境，在这种环境中，人们特别关注物质的直接再现及其触觉效果，其中的亲切感是现实中多次突变的结果，它成为与观众建立联系的不可或缺的要素。

Close Up Print 特写

Prints 版画

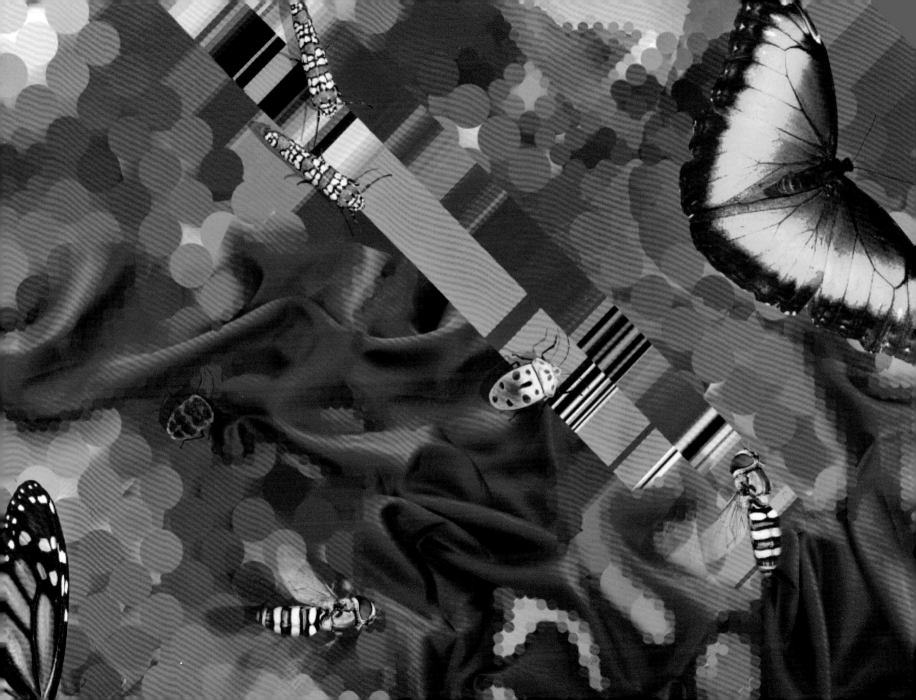

EYE CANDY
PACIFIC DESIGN CENTER

Date: 2013

Client: Industry Gallery

Location: Pacific Design Center in Los Angeles, California, USA

秀色可餐
太平洋设计中心

时间：2013 年

客户：工业画廊

地点：美国，加利福尼亚州，洛杉矶太平洋设计中心

"Eye Candy" is a site-specific installation for the Pacific Design Center in Los Angeles. The work explores the cultural and phenomenological potential of synthetic surface finishes at architectural scale. The unique iconography of this installation puts forth a clear statement that architectural material finishes have the communicative potential to enter into the imaginary realm of our 'eye-candy' culture, exploiting our most superficial of obsessions surrounding desire, age, gender, media, consumption, and delight.

This body of work turns the traditional cliché of architecture inside out: rather than embodying an eternal adult essence, the graphic subject of 'eye-candy' iconography is the result of superficial mass-culture desires, allowing the visitors to indulge in their infant fantasies. The chromatic gradients range from shades of pink to light green, in a high gloss metallic finish. Initially, 3D scanned candies are used as a new geometrical matter than can be scripted into stripes of gradient colorations. The candy configuration strives for the abstraction and flatness typical of a cartoon fantasy, and diverges from any manner of 'truth' in realism and representation.

"秀色可餐"是洛杉矶太平洋设计中心的特定场地装置。该装置探索了应用于建筑学领域的合成材料在文化和现象学方面的影响。这个设计独特的图像清楚地表明，建筑合成材料具有交际功能，能够将建筑设计提升到令人大饱眼福的新高度，同时也能够发掘不同年龄、性别、媒介以及消费层次的人的不同喜好。

该作品颠覆了人们对建筑的传统观念，它不再注重体现人类永恒的本质，它的形象是大众文化欲望的结果，可以让游客沉溺于他们少年时期的幻想。作品的色彩范围从粉红色到浅绿色，并且用高光泽的金属色构建完成。三维扫描糖果被用作一种新的几何物质，被脚本化成渐变颜色的条纹。糖果的造型力求抽象和平面化，这是卡通的典型特征，不同于任何现实主义的表现形式。

Conceptual Drawing 概念图

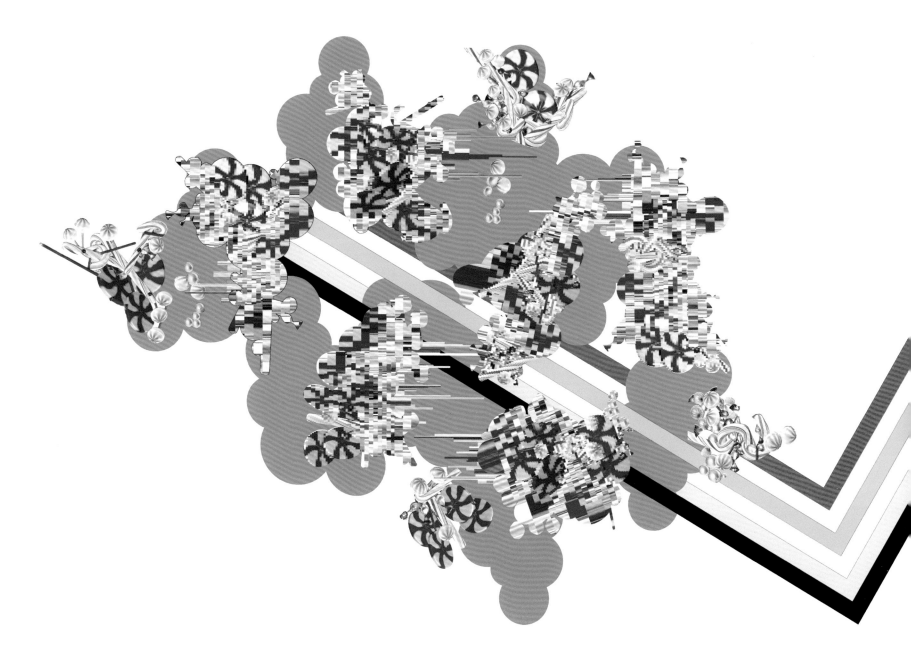

Elevation 正面

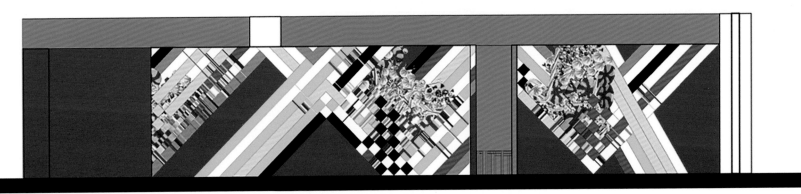

Elevation 正面

Today, beauty is not a singular idea, its plurality prevails. It is the creative role of the architect to bring profound new architectural beauties to cities, substituting alienation with widening the palate of our emotional involvement, and introduce contemporary and diverse experiences of beauty into architecture. This is a concept for today's age, a product of individuals for other individuals, authentically celebrating pluralism.

Yael Reisner

如今，美不再是一个单一的概念，它的多元性开始盛行。建筑师的创造性设计为城市带来全新而深刻的建筑美，用扩大情感介入的范围来取代疏离，并将当代多元素美感的体验引入建筑。这是当今时代的一个概念，是个体为其他个体创造的产物，是真正意义上的多元主义的胜利。

雅尔·雷斯纳

Benches 长椅

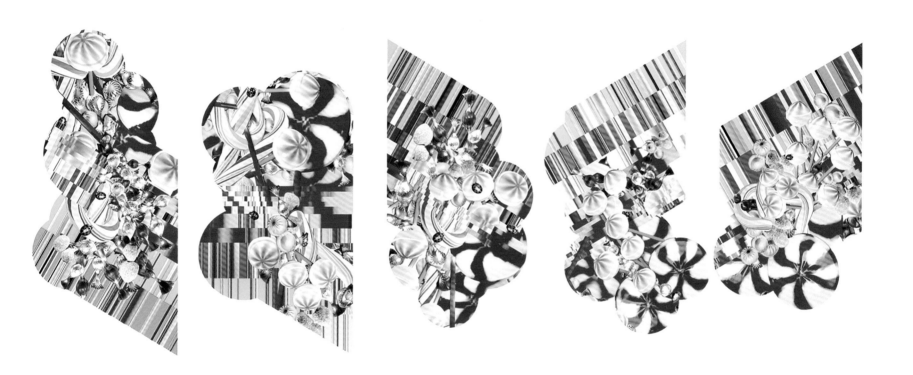

TEMPERA
MOCA

Date: 2012
Client: Museum of Contemporary Art
Location: Los Angeles, California, USA

彩画
洛杉矶当代艺术博物馆

时间：2012 年
客户：当代艺术博物馆
地点：美国，加利福尼亚州，洛杉矶

"Tempera" is an indoor pavilion design proposal commissioned by the Museum of Contemporary Art of Los Angeles, for a show focused on Southern California architecture after 1987, and specifically on Los Angeles as a natural location for architectural innovation. The design creates a fantastic garden where the visitors are able to see their own images reflected into a three-dimensional immersive painted canvas.

The massing of the pavilion is a tilted cube. The simple act of tilting a primitive volume has the strength to challenge the idea of the ground and the stable relationship between architecture, the horizontal plane, and the viewers.

The outside surface of the pavilion is covered by 150 floral panels; each floral panel is made of folded aluminum, powder coated, and assembled onto a modular triangular structure. The flower configuration, relentlessly arranged in repeated panels, strives for an abstraction and flatness that diverges from any attempts at realism.

"彩画"是受洛杉矶当代艺术博物馆委托的一个室内展馆设计方案，旨在展示1987年后南加州的建筑，特别是展现洛杉矶作为建筑创新的自然场所。该设计创造了一个奇妙的花园，游客可以看到自己的形象映射在一个三维沉浸式油画画布中。

展馆的外形是一个倾斜的立方体。之所以把它设计成倾斜的就是要挑战建筑、水平面和参观者之间趋于稳定的关系。

展馆的外表面由150个花卉面板覆盖，每个花卉面板由折叠铝制成，粉末涂层，并组装到模块化的三角形结构上。花朵反复排列在重复的面板中，不同于任何现实主义的尝试，它力求抽象和单调。

Installation 安装

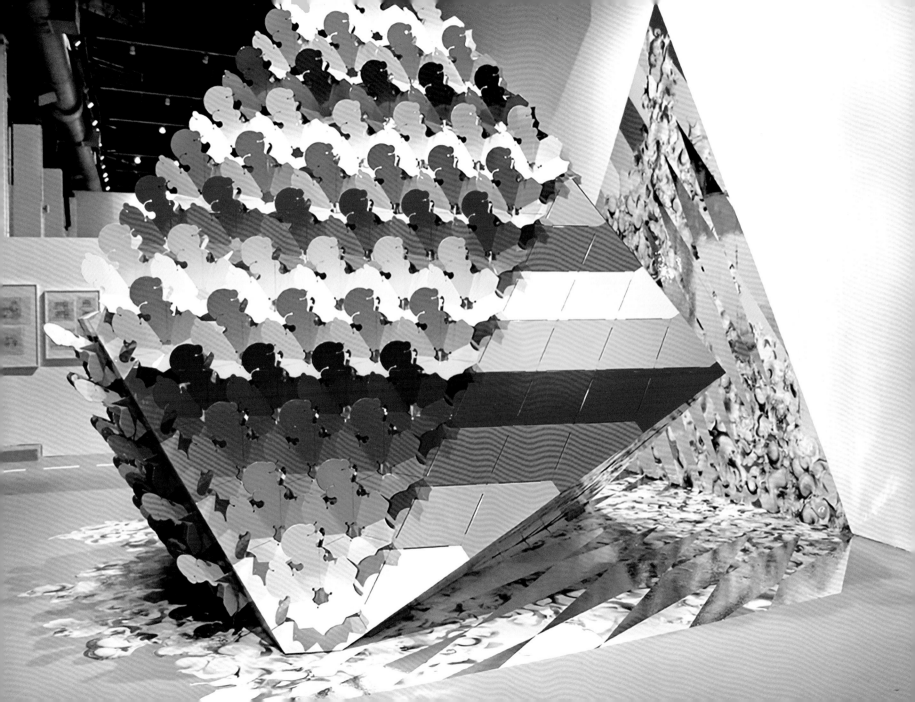

Elevation 正面

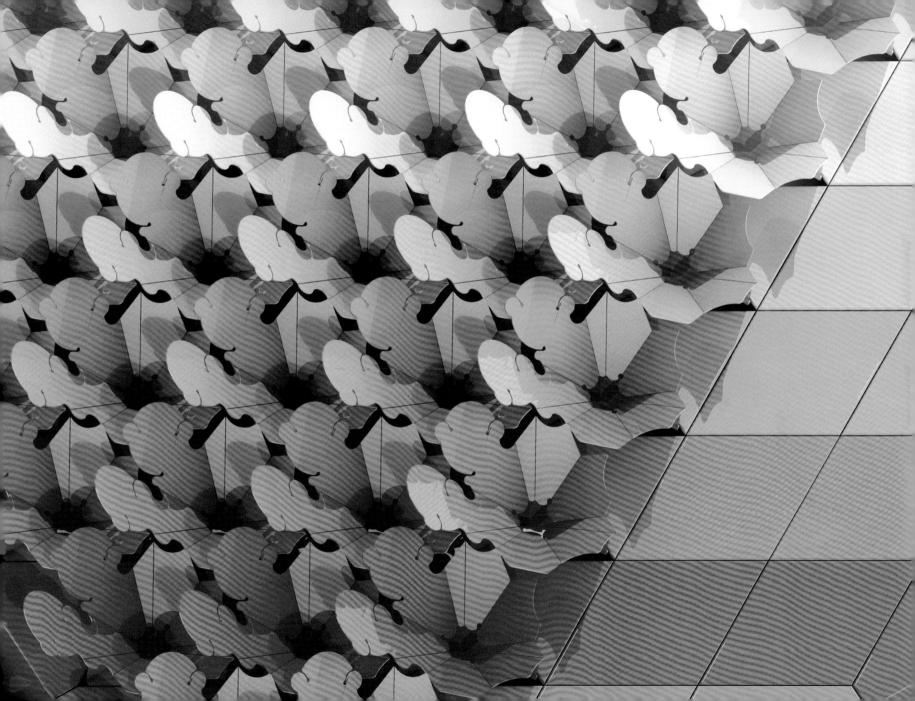

Architects are experts at being generalists. We are always multidisciplinary. In the last thirty years this has been accelerating exponentially due to the technologies and platforms for the distribution of knowledge and of information. These platforms are the same for multiple cultures of design.

Hernan Diaz Alonso

建筑师是综合型人才。我们要实现多学科的融合。在过去的30年里,随着知识和信息传播技术以及平台的出现,融合的数量呈指数级增长,多元文化设计也随之发展了起来。

埃尔南·迪亚兹·阿隆索

Details 细节

Details 细节

The phenomenon known as Beauty, since ancient times until today, arouses and stimulates human emotions; triggering elation or melancholy by taking us out of being indifferent or passive. There are ways in which space becomes emotionally charged.

There is a need for a personal call to arrive at new beauties. Poetic architecture, as poetry, evolves from a private psyche, though, as G. Bachelard wrote, "the poet does not confer the past of his image upon me, and yet his image immediately takes root in me."

Yael Reisner

从古至今,那些被称为美的景象,唤起和刺激了人类的情感,使我们从冷漠消极中解脱出来,让我们感受到愉快或抑郁,快乐或悲伤。有很多设计方法可以让空间充满感情。

你需要自己去感受才能获得美的景象。诗意建筑,正如诗歌一样,从个人心灵中进化而来,源自一个人的灵魂深处,像巴切拉德所写,"诗人没有把他过去的形象赋予我,但他的形象却能立即在我心中扎根。"

雅尔·雷斯纳

Details 细节

Installation Details 安装细节

Installation Details 安装细节

INVERTED LANDSCAPES

Zev Yaroslavsky Family Support Center Civic Artwork

Date: 2016

Client: LA County Art Commission

Location: San Fernando Valley, California, USA

"Inverted Landscapes" is a set of two public artworks for the San Fernando Valley Family Support Center atrium and garden. Nature is a silent but powerful protagonist of both art pieces. Through the use of highly photorealistic depictions as a point of departure, the two expand on the notion of contemporary paintings and update viewer's classical notion of two-dimensionality.

"Inverted Landscapes" implies that everything is relative to one's vision of the surrounding reality, and that often being able to shift one's point of view is the first step to understanding the people around us. The hung ceiling in the lobby portrays a highly photorealistic representation of the LAND, which, although firmly anchored to the ground, revolves around an abstract visualization of the SKY. The outdoor art-proposal is constructed by a series of stable and interrelated strips, that move from the ground onto the glass façade of the building. The depiction of the sky not only implies a sense of hope, but most importantly gives a sense of perspective to one's approach to life.

倒置景观

圣费尔南多家庭支持中心

时间：2016年

客户：洛杉矶县艺术委员会

地点：美国，加利福尼亚州，圣费尔南多山谷

"倒置景观"是由埃琳娜·曼费迪尼设计的位于圣费尔南多家庭支持中心前厅和花园的两件公共艺术作品。自然是这两件艺术作品中沉默而有力的主角。两件作品以高度写实的绘画为出发点，拓展了当代绘画的概念，更新了观者的经典二维观念。

"倒置景观"意味着一切想法都是一个人对于周围现实世界的看法，而改变自己的看法是我们理解其他人的第一步。大厅的吊顶高度逼真地描绘了土地的景象，虽然牢牢地固定在地面上，虽然描绘的是大地景象，但周围却环绕着抽象的天空景象。户外艺术方案由一系列稳定的、相互关联的条带构成，这些条带从地面移动到建筑的玻璃立面上。对天空的描绘不仅意味着希望，更重要的是提供给人一种全新的看待生活的视角。

Art Glass for Building Curtain Wall
建筑幕墙的艺术玻璃

"Civilizations could not exist without the recurrence of pleasures, including experiencing beauty", claims Semir Zeki, the Neurobiologist and Prof. of Neuroesthetics, who shows through his findings, that this dependency is part of our neurobiological structure, and therefore, that daily pleasure makes people not only happier but also healthier.

The experience of beauty rewards people neurobiologically with an immediate reaction of an aesthetic pleasure, which affects physical health as well as a feeling of wellbeing. Zeki found that when people look at objects they consider to be beautiful, there is an increased activity in the pleasure reward centers of their brain. There is a great deal of dopamine in this area, also known as the "feel-good" transmitter. "The reaction is immediate." The experience of beauty is instantly recognizable through the intensity of that emotional experience, which can be quantified digitally.

Yael Reisner

神经生物学家塞莫·泽克教授声称:"没有快乐的重现,文明和美的体验不可能存在。"他的发现显示,这种相依性是我们的神经结构的一部分,因此,日常的愉悦不仅使人更加快乐,还会更加健康。

从神经生物学的角度来说,美的体验会给人带来审美愉悦的即时反应,从而影响身体健康和幸福感。泽克发现,当人们看到他们认为漂亮的东西时,他们大脑中的快乐激励区域的活动会增加。这个区域可分泌大量的多巴胺,而多巴胺也被称为"感觉良好"的递质。人们的反应是迅速的,"通过强烈的情感体验,人们可以立即识别出其中的美的体验,而这种情感体验可以用数字量化。"

雅尔·雷斯纳

Digital Graphic 数字图形

Interior View 内部景观

Exterior View 外部景观

Suspended Ceiling 吊顶

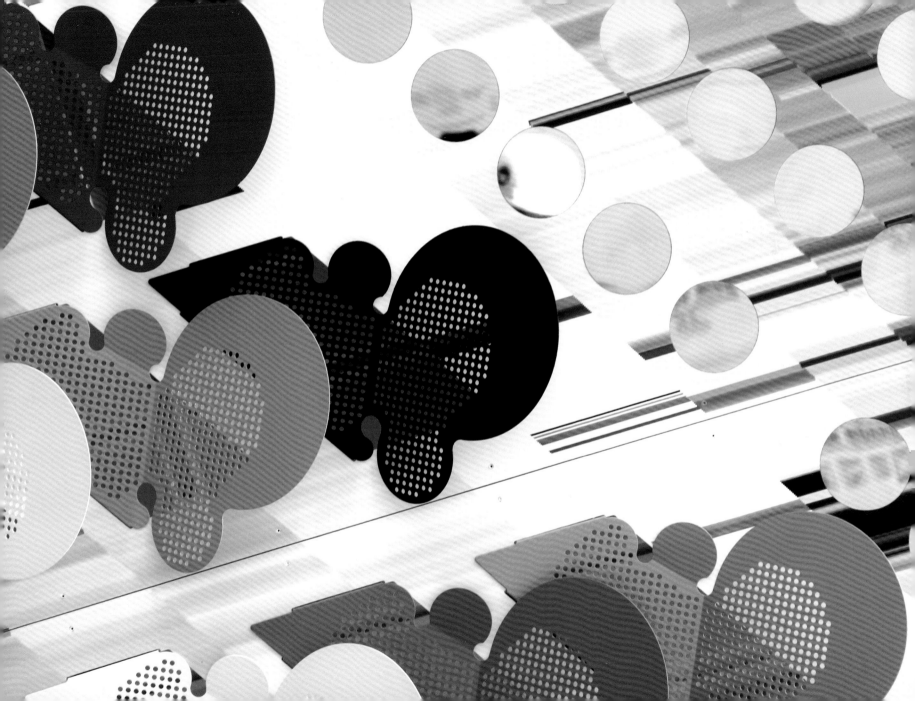

LIVING PICTURE

KAIDA CENTER OF SCIENCE AND DESIGN

Date: 2019

Client: Kaida

Location: Dongguan, China

生活图片

凯达科学与设计中心

时间：2019 年

客户：凯达

地点：中国，东莞

Situated in a lush ecosystem, the Kaida Center of Science and Design has the opportunity to participate in the dialogue between urban development and landscape preservation in the fast growing city of Dongguan, China. Oftentimes as urban centers develop, they forget their connection to the ecological environment, and beautiful green landscapes are replaced with concrete and mortar.

Contrary to many other urban developments, the Kaida Ecological Office District's intention is to bring nature into the city. Using this as a point of departure, Atelier Manferdini's concept "Living Picture" makes inhabitants and visitors conscious towards nature. The artwork has three components: a colorful immersive art wall, a pair of connective circulation systems, and high resolution printed glass. From its inception, the artwork was developed with the intention of connecting the natural with the digital. This 5.5 meter tall art wall transforms the connective corridor into an immersive environment for viewers to get lost in. The upper portion is composed of a fixed metal tile mosaic that depicts a color environment while the lower 2 meters is composed of 1,700 rotating blocks that encourage the spectator to engage with the artwork. As the viewer rotates the lower blocks, they leave their own presence on the artwork.

凯达科学与设计中心位于郁郁葱葱的生态区内，它有机会在中国快速发展的城市东莞参与城市发展和景观保护的"对话"。随着城市中心区域的发展，人们忘记了建筑与生态环境之间的联系，美丽的绿色景观被混凝土和砂浆所取代。

与其他许多城市发展相反，凯达生态办公区的设计理念是将自然带入城市。以此为出发点，曼费迪尼事务所的"生活图片"概念使居民和游客产生对自然的思考。该艺术装置有三个组成部分：彩色的沉浸式艺术墙、一对连接的循环系统、高分辨率的印刷玻璃。从一开始，作品就以连接自然和数字为目的。这座5.5米高的艺术墙将走廊变成了一个沉浸式的环境，让观众沉浸其中。它的上半部分由固定的马赛克金属瓷砖组成，描绘了一个彩色的环境，而下面2米由1700个旋转方块组成，从而鼓励观众参与到艺术作品当中。当观众旋转较低位置的方块时，他们便会在艺术品上留下自己的记号。

Interactive Wall 互动墙

In 2011, Zeki proved that humans experience four types of beauty in their emotional brain — visual beauty, musical beauty, moral beauty, and mathematical beauty, and that each can be recorded and quantified.

Nonetheless, there is no set of rules on how to create beautiful artifacts, art or architecture, and as Zeki confirms, no one can define beauty in simple terms. Mathematicians — who never stopped believing in the role of beauty, unlike architects—who are knowingly, and comfortably, able to list characteristics such as surprise, significance, clarity, profundity, or ambiguity, and when order takes over disorder, "all falls into place", a pleasing well known step towards beauty.

Elena Manferdini's work is typified by all the characteristics listed here, with an emphasis on profundity, as most architects don't associate beauty with it. It is clear how she creates a tighter affinity between Nature and Culture, and very much through the aesthetics of her projects.

Yael Reisner

2011年，泽克证明了人类情感可以体验到的四种美——视觉美、音乐美、道德美和数学美，每一种都可以被记录和量化。

尽管如此，他却没有阐释如何创造关于物品、艺术或建筑的美的规则，因为正如泽克所确认的那样，没有人可以用简单的术语来定义美。数学家们一直坚信美的作用，建筑师们却会有意识地、随意地、舒适地列出诸如惊人、重要、清晰、深刻或模糊等特征，当秩序取代混乱时，便"一切就绪"，迈出美丽的、令人愉快的、众所周知的一步，自然就创造出了美。

埃琳娜·曼费迪尼的作品以这里列出的所有特征为代表，强调了深度，因为大多数建筑师并不把美与它联系起来。很明显，通过她的作品，自然和人类文明出现了一种更紧密的联系。她是如何在自然和文化之间创造出更紧密的亲近感的呢？在很大程度上是通过她作品中的美来实现的。

雅尔·雷斯纳

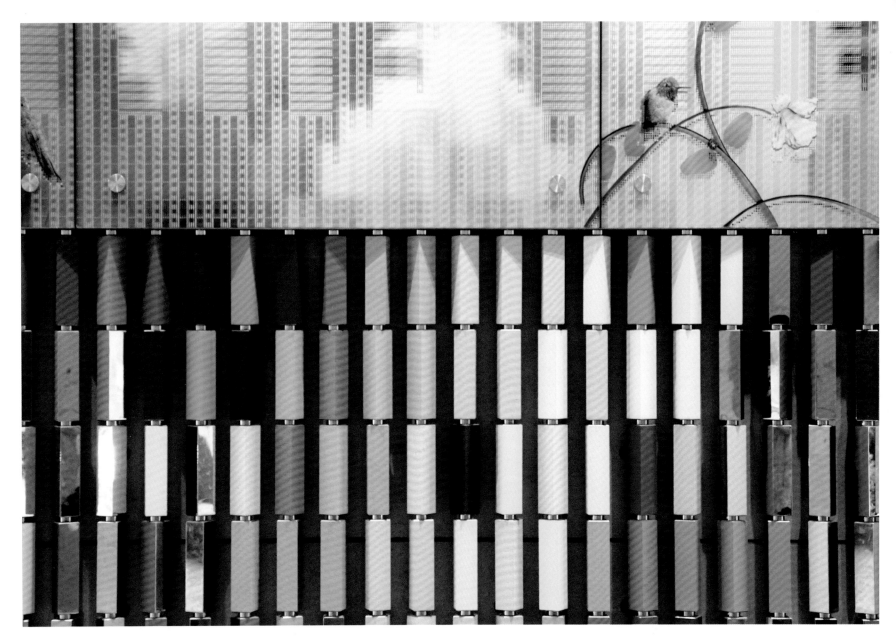

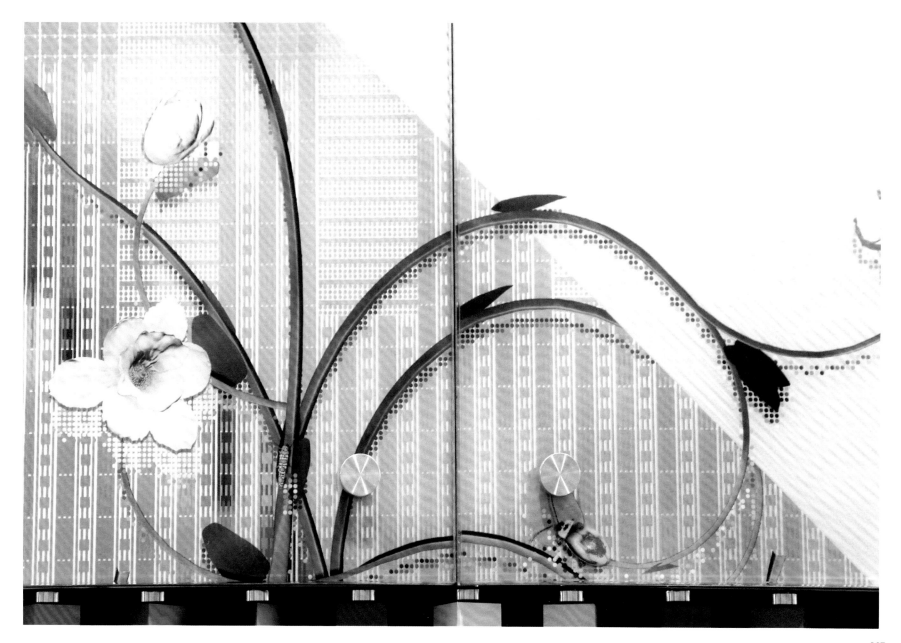

Close Up Print 特写

More architects need to follow Elena Manferdini in her ability to avoid holding back, or denying lateral thinking, intuition, speculation, "irrational" associations, memories and personal projections while producing poetic and mental images along with, and as part of, anything else in the complex process of making Architecture. This is a vital decision if one is eager to get involved intentionally with aesthetics, and intended to affect. Often, personal and intimate landscapes reflect on the collective culture and becomes a registry of the culture in this time. This is necessary, not only to bring cultural meaning locally, but also to attain cultural diversity in architecture by nurturing and preserving the variance amongst individuals.

Yael Reisner

更多的建筑师需要遵循埃琳娜·曼费迪尼的理念：避免设计上的有所保留，要打破横向思维、直觉、猜测、"非理性"的联想、记忆和个人感情投射，在建筑设计的复杂过程中产生诗意和心理意象，并成为其他任何东西的一部分。

如果一个人有意识地从美学的角度参与其中并且试图造成影响，那么这将是一个至关重要的因素。通常，相对私人化的景观却反映了群体文化并成为当时文化的印记。这是非常必要的，不仅赋予了当地的文化含义，还通过促进和保留个体之间的差异实现了建筑的文化多样性。

雅尔·雷斯纳

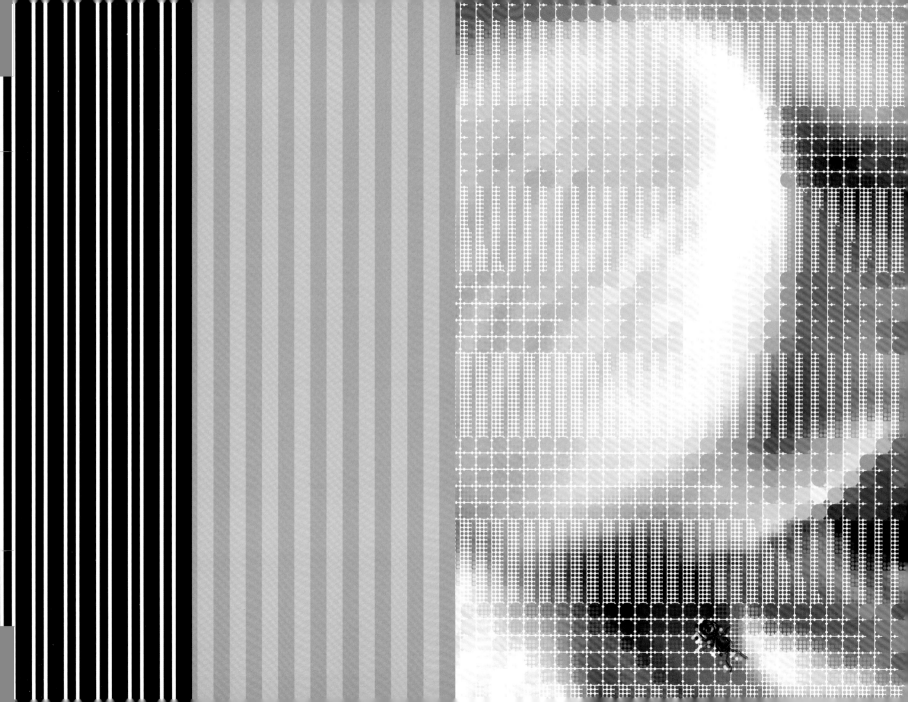

IMAGE ANNOTATION
图片注释

LANE CRAWFORD 连卡佛

017 Blueberries | Lane Crawford | 2018
蓝莓 | 连卡佛 | 2018 年

018 Daisy | Lane Crawford | 2018
雏菊 | 连卡佛 | 2018 年

021 Folds | Lane Crawford | 2018
折叠 | 连卡佛 | 2018 年

022 Delphinium | Lane Crawford | 2018
飞燕 | 连卡佛 | 2018 年

023 Folds | Lane Crawford | 2018
折叠 | 连卡佛 | 2018 年

025 Folds | Lane Crawford | 2018
折叠 | 连卡佛 | 2018 年

027 Folds | Lane Crawford | 2018
折叠 | 连卡佛 | 2018 年

029 Folds | Lane Crawford | 2018
折叠 | 连卡佛 | 2018 年

032 Shanghai Window | Lane Crawford | 2018
上海橱窗 | 连卡佛 | 2018 年

034 Shanghai Window | Lane Crawford | 2018
上海橱窗 | 连卡佛 | 2018 年

038 Shanghai Window | Lane Crawford | 2018
上海橱窗 | 连卡佛 | 2018 年

041 Beijing Window | Lane Crawford | 2018
北京橱窗 | 连卡佛 | 2018 年

URBAN FABRIC RUG 城市肌理地毯

045 Building Portraits Rugs | 2018
建筑肖像地毯 | 2018 年

047 Building Portraits Rugs | 2018 | LACMA permanent collection
建筑肖像地毯 | 2018 年 | 洛杉矶博物馆永久收藏

050 Silk Textiles | Building Portraits Rugs | 2018
丝绸纺织品 | 建筑肖像地毯 | 2018 年

FRAMED SKYLINES 天际轮廓线

055 Framed Skylines | Metro | 2016
天际轮廓线 | 地铁 | 2016 年

056 Elevation | Metro | 2016
正面 | 地铁 | 2016 年

059 Diagrams | Metro | 2016
分析图 | 地铁 | 2016 年

LOO WITH A VIEW 观景洗手间

061   Digital Print | Longmont Museum | 2015
      数字印刷 | 朗蒙特博物馆 | 2015 年

063   Elevation and Reflected Ceiling | Longmont Museum | 2015
      正面和天花板 | 朗蒙特博物馆 | 2015 年

064   Installed Piece | Longmont Museum | 2015
      安装件 | 朗蒙特博物馆 | 2015 年

MASSIVE PROJECTIONS 巨幕投影

069   Wallpaper | 5th Gwanjiu Design Biennial | 2013
      壁画 | 第五届韩国光州设计双年展 | 2013 年

071   Wallpaper | 5th Gwanjiu Design Biennial | 2013
      壁画 | 第五届韩国光州设计双年展 | 2013 年

FRESCO 壁画

075   Close Up Printed Vinyl | Design Matters Gallery | 2012
      特写玻璃纸 | 设计画廊 | 2012 年

076   Digital Print | Design Matters Gallery | 2012
      数字印刷 | 设计画廊 | 2012 年

FABULATIONS 虚构情节

081   Close Up Print | Istanbul Design Biennial | 2014
      特写 | 伊斯坦布尔设计双年展 | 2014 年

084   Prints | Istanbul Design Biennial | 2014
      版画 | 伊斯坦布尔设计双年展 | 2014 年

EYE CANDY 秀色可餐

089   Conceptual Drawing | Pacific Design Center | 2013
      概念图 | 太平洋设计中心 | 2013 年

090   Elevation | Pacific Design Center | 2013
      正面 | 太平洋设计中心 | 2013 年

092   Elevation | Pacific Design Center | 2013
      正面 | 太平洋设计中心 | 2013 年

093   Benches | Pacific Design Center | 2013
      长椅 | 太平洋设计中心 | 2013 年

TEMPERA 彩画

097   Installation | MOCA | 2012
      安装 | 洛杉矶当代艺术博物馆 | 2012 年

098	Elevation I MOCA I 2012 正面 I 洛杉矶当代艺术博物馆 I 2012 年	115	Digital Graphic I Zev Yaroslavsky Family Support Center Civic Artwork I 2016 数字图形 I 圣费尔南多家庭支持中心作品 I 2016 年
101	Details I MOCA I 2012 细节 I 洛杉矶当代艺术博物馆 I 2012 年	118	Interior View I Zev Yaroslavsky Family Support Center Civic Artwork I 2016 内部景观 I 圣费尔南多家庭支持中心作品 I 2016 年
102	Details I MOCA I 2012 细节 I 洛杉矶当代艺术博物馆 I 2012 年	120	Exterior View I Zev Yaroslavsky Family Support Center Civic Artwork I 2016 外部景观 I 圣费尔南多家庭支持中心作品 I 2016 年
105	Details I MOCA I 2012 细节 I 洛杉矶当代艺术博物馆 I 2012 年	122	Suspended Ceiling I Zev Yaroslavsky Family Support Center Civic Artwork I 2016 吊顶 I 圣费尔南多家庭支持中心作品 I 2016 年
106	Installation Details I MOCA I 2012 安装细节 I 洛杉矶当代艺术博物馆 I 2012 年		
108	Installation Details I MOCA I 2012 安装细节 I 洛杉矶当代艺术博物馆 I 2012 年		

INVERTED LANDSCAPES 倒置景观

LIVING PICTURE 生活图片

113	Art Glass for Building Curtain Wall I Zev Yaroslavsky Family Support Center Civic Artwork I 2016 建筑幕墙的艺术玻璃 I 圣费尔南多家庭支持中心作品 I 2016 年	129	Interactive Wall I Kaida Center of Science and Design I 2019 互动墙 I 凯达科学与设计中心 I 2019 年
		138	Close Up Print I Kaida Center of Science and Design I 2019 特写 I 凯达科学与设计中心 I 2019 年

埃琳娜·曼费迪尼作品解析（下）

PORTRAITS

竖直

李宁　[意]埃琳娜·曼费迪尼　著

江苏凤凰科学技术出版社

CONTENTS
目录

4	ON LINES AND THEIR WEIGHTS	关于线和线宽
8	BUILDING THE PICTURE	构建图片
38	BUILDING PORTRAITS	建筑肖像
58	INK ON MIRROR	镜子上的墨水
98	TWONESS	两元一体
108	CHICAGO SKYLINE	芝加哥天际线
112	AYALA	阿亚拉
128	AT HUMAN SCALE	人体尺度
136	BLANK FACADE	空白立面
140	ALEXANDER MONTESSORI SCHOOL	亚历山大蒙特梭利学校
148	CABINET OF WONDERS	神奇陈列柜
156	WOVEN HOUSE	编织房屋
166	BIOGRAPHY	作者介绍
168	CONTRIBUTORS	贡献者
172	IMAGE ANNOTATION	图片注释

ON LINES AND THEIR WEIGHTS

BY JASMINE BENYAMIN

To draw a line is to have an idea. Ideas become compounded as soon as you make the second one.

—Richard Serra[1]

Try always, when you look at a form, to see the lines in it, which have had power over its past fate, and will have power over its futurity.

—John Ruskin[2]

Where there is architecture, there are lines.

In his studies for an early painting "Adoration of the Magi," Leonardo da Vinci utilized the recently inaugurated mathematically derived technique of one point perspective to stage his figures in a scene that notably included architectural ruins. Of the extensive line work in these and other drawings, the most consequential were those that formed the Cartesian grid converging at a designated horizon line. Even though they were subsumed by paint in their final iteration, they were crucial in determining the structuring of the picture plane. In Albertian terms, Leonardo was designing the space of action. Lineamenta before materia.

The primacy of line work in architectural delineation arguably proceeds from this moment. Intrinsic and extrinsic, hidden and revealed, sharp and blurred – lines have literally and figuratively ruled the way in which we manifest ideas onto the physical world. Plans, sections and elevations make up the arena where lines index a whole range of conventions that constitute a kind of code. Once the rules are learned, they are hard to unlearn, let alone un-see.

Recently, lines have taken a curious – but perhaps necessary – turn. As modes of digital drawing have proliferated, it appears as if anything goes. Orthographic projection – those drawings that have governed pedagogy on representation in architecture schools since the beginnings of the discipline – no longer hold a dominant place in the discourse. Lines that had hitherto been reliable witnesses to tectonic and spatial narratives are no longer stable.

The old story reads something like this: in order to make drawings 'legible,' one needs to control how to represent what is cut and what is not; thicker for the former, thinner for the latter. Double thick lines indicate a wall, dashed lines for what may be moving or what is above. A

---

1. Richard Serra, *Writings, Interiors* (Chicago: University of Chicago Press, 1994), p. 52.
2. John Ruskin, *The Elements of Drawing*, (NewYork: Dover, 1971), p. 91.
3. As Mario Carpo puts it, Alberti defined perspective drawings as light rays traced on a surface. Mario Carpo, *The Alphabet and the Algorithm*, (Cambridge: MIT Press, 2011), p. 12.

range of other line types are deployed to designate material and surface properties; hatches, dots, curves, etc. are clues to a bigger story. As for so called construction lines – fetishized or not – no manually drafted projection would be possible without them. Before all these elements can be drawn, a matrix of ghostly rays[3] needs to be established. They are not there to be pretty, but to make sense of things. They lie under and between shears and overlays, and the fact that they extend at times beyond their relevance is besides their point.

It should be noted that these lines continue to proliferate in practice, but if one were to see where things are getting interesting, the academy is a good place to look. Schools are presenting students with ever expanding toolkits that challenge Alberti's delineation of ideation and crafting, with what Mario Carpo coins "digital artisanship."[4] Enter Elena Manferdini, a pedagogue and practitioner whose work arguably reconciles these growing discursive breaks between the discipline and the profession. Binaries of past and present thinking on lines– analog/digital, representation/simulation, abstraction/concreteness, and imagination/reality– don't apply. In a recent exhibition at the A and D Museum in Los Angeles entitled "A Line of Inquiry," Manferdini stated that the works on view performed a "tension between intuition and reflection, between concreteness and abstraction", and that they toggled "between being a way to imagine a plausible reality, or simply as a means to an end."[5]

Do these artifacts precede or post-rationalize objects? In the same introductory statement, she refers to the works on view as 'elevations' that are 'autonomous.' Indeed, to say that they operate as either part of a larger set of documents or stand alone statements is to oversimplify. After all, the drawings straddle both architecture and art, so we are free to speculate on the origins and contexts of meaning production.

It would also be 'old school' to dismiss these images as scale-less and site-less; the compositions of lines of various thicknesses and colors operate at both local (gradient) and global (field) registers. This is true of both the drawings that have a bottom or ground, and those that appear to be constrained only by the edges of media (paper, metal, vinyl). Others are clearly meant to be read as axonometric, and still others as cut and paste fragments. Alongside the accumulation of line work, there are strokes and marks, and in an explicit nod to their digital disposition, swaths and patches of low-resolution pixels. There is an unrelenting linearity about everything: when subject to a squint test, axiality and movement vectors prevail. The formal operation they share lies in the concept of accretion. More is more.

Some of the drawings scale up to physical realization as free-standing sculpture, screens, and facades. Within the space of the gallery, these lines also delineate extents that fold over floors and up walls. The fact that the work is also now being fabricated as textiles appears to be a natural leap, since the scripted flatness of the drawings nonetheless telegraphs an abundance of weaving and layering. The drawings want to be enacted in other spatial scenarios.

It would be disingenuous to say that I know exactly how Manferdini's drawings are produced. I have some idea, but I must rely primarily on what I see. They are equal parts mysterious and revelatory. Whether as edges, borders, or limits, as indexes of material change or shifts in planes of extrusion, as arcs or bends of a curve, her lines are meant to be viewed, read and inhabited. The possibilities are endless.

---

4. Mario Carpo, *The Alphabet and the Algorithm*, (Cambridge: MIT Press, 2011), p. 117.
5. The exhibition was on view between 16 February 2018 and 20 April 2018.

关于线和线宽

杰思敏·本雅明

一根线条是一种思维的映射，而第二根线条的出现则是思维整合的结果。

——理查德·塞拉[1]

请学会尝试从一张表格中看到那些维系着从过往命运连接并影响着未来的线索，即那些线条。

——约翰·拉斯金[2]

建筑即是线条所被赋予的具象化。

在达·芬奇的早期绘画作品"三博士来朝"中，他使用了全新开创性数学式的单点透视法，在一个包含了建筑遗址的场景中表现画中的人物。在这幅画以及其他作品大量出现的线条中，最为重要的是形成了交会于视平线上的笛卡尔网络的那些线条。尽管这些线条在作品最终完成后被涂料覆盖，但它们对于确定画面结构起到了至关重要的作用。用爱因斯坦的话来说：达·芬奇是在设计行为活动的空间（物质性之前的线性）。

线条在建筑描绘中的首要地位的确定可以说是从这一刻开始的。内在的和外在的，隐藏的和暴露的，尖锐的和模糊的——线条从字面上和形象上统治着我们在物质世界中表达思想的方式。在平面图、剖面图和立面图所构成的舞台上，线条记录了一系列由惯例所构成的规则，这种规则一经学习便很难遗忘，更不可能被视而不见。

近代以来，关于线条的研究发生了一个有趣的（但可能是必要的）变化。随着数字绘图模式的兴起，曾经的一切似乎都不复存在。自建筑学科伊始就一直支配着建筑表现教学的正交投影法，不再占据主导性话语权。迄今为止，仅靠线条绘制构造和叙述空间的方法已不再适用了。

传统的建筑绘图方法为了使图纸"清晰易读"，需要用线条来表示被遮挡的和不被遮挡的物体；越靠前的物体线条越粗，靠后的则越细；双粗线表示墙，虚线表示可能移动的内容或覆盖

1.理查德·塞拉.室内设计[M].芝加哥：芝加哥大学出版社，1994：52.
2.约翰·拉斯金.绘画的元素[M].纽约：多佛，1971：91.

在上方的内容。建筑师还使用了一系列其他线条类型来指定材质和表面属性。阴影、点线、曲线等用来表达更大范围的内容。至于所谓的辅助线——不管是否被建筑师所信赖——没有它们，任何手工绘制的投影都是不可能的。在绘制所有这些元素之前，一般还需要建立一个隐形的射线矩阵[3]。不是为了好看，而是为了使图面内容更加合理。它们位于不同的图层之间或者背景层，有时甚至超出了与它们相关联的内容。

值得注意的是，这些线条在实际操作中不断增多，如果想要了解该艺术形态的变化，需要从学术角度分析。学校为学生们提供了时时更新的工具集，用以挑战爱因斯坦对思维和工艺的描绘，马里奥·卡波称之为"数字艺术"[4]。教育学家和艺术从业者埃琳娜·曼费迪尼的作品调和了这些学科和专业之间不断扩大的争论。过去和现在对于线条产生思考的完全对立的形态——模拟/数字、表现/模拟、抽象/具体以及想象/现实——均不再适用。在最近洛杉矶建筑与艺术博物馆举办的名为"探究线条"的展览中，曼费迪尼表示，展出作品表现出"直觉和反思之间、具体和抽象之间的张力"，它们在"想象一个可信的、现实的或者简单作为一种达到目的的手段"这两种方式之间切换[5]。

曼费迪尼的这些作品是在形成实体之前还是之后被合理化的？在曼费迪尼为本书所写的简介中，她将展出作品称为"独立"的"立面"。事实上，说它们是作品的一部分或杰出论述的一部分，都显得过于简单化。毕竟，这些作品跨越了建筑和艺术，我们可以自由地推测其意义的起源和产生背景。

将这些图像视为"无比例""无文脉"也是保守的做法，不同厚度和颜色的线条的组合在局部（梯度）和全局（场）都发挥着作用。这对于有底图或底纹的图纸，以及那些看起来仅受介质边缘（纸张、金属、乙烯基）约束的图纸均适用。伴随着线条的各种组合叠加，图像中还有加粗和标记，为了精确的模拟数字版本，还包含条带和低分辨率线条的平铺。每幅作品都具有持续的线性特征：当对画作进行斜视观测时，轴性和运动矢量随处可见。它们共同的形式化的作用潜藏于累积性的理念中，积累越多则越丰富。

一些图纸回归到现实中会变为独立的雕塑、屏风和建筑立面。在画廊的空间内，这些线条还勾勒出地板和墙。这些作品被当作纺织品来制作似乎是一个很自然的跨越，因为图纸所描绘的平面也显示出大量的线条交织和层次感，这些画作需要在其他空间场景中被展现出来。

说我完全了解曼费迪尼的作品制作过程并不现实，我只是有些许了解，我主要依靠我所看到的事实来推理。它们神秘而同样充满启发，无论是作为边缘、边界或界限，还是作为物质在延展的平面中产生的变化或移动，抑或曲线的弧度或弯曲的形式，她的线条都值得被欣赏、解读和定义，这里面蕴含着无限的可能。

---

3. 正如马里奥·卡波所说，阿尔伯蒂将透视图定义为在表面上描绘的光线。马里奥·卡波.字母表和算法[M].剑桥：麻省理工学院出版社，2011：12.。
4. 马里奥·卡波.字母表和算法[M].剑桥：麻省理工学院出版社，2011：117.。
5. 该展览于2018年2月16日至2018年4月20日期间展出。

BUILDING THE PICTURE
ART INSTITUTE OF CHICAGO

Date: 2015
Curator: Zoe Ryan
Location: AIC, Chicago, Illinois, USA

构建图片
芝加哥艺术博物馆

时间：2015年
策展人：佐伊·瑞恩
地点：美国，伊利诺伊州，芝加哥，芝加哥艺术博物馆

To any outsider observing architecture, and specifically to architects, it might appear obvious that this field has seemingly unlimited faith in the power of geometry. One such geometry—perhaps the most diffused one—is the grid: an overall mastering system from which the professional trade has been able to lock into place a good portion of our built environment. Modern American cities stand as witnesses to this need to represent analytically and graphically the rigor of a global organizational rule which we can all comprehend.

This exhibition is part of a series in which the Department of Architecture and Design enlists contemporary architects and designers to organize installations that investigate critical issues within their practices. Manferdini began with Mies's simple gridded facade treatments. After tracing an image of his facades to create digital drawings of the grid, she developed her own unique skyline for the city of Chicago. Manferdini's manipulation of the grid blurs the line between fashion and pattern in an architectural context and introduces a new contemporary landscape that has strong ties to the past. Using history as a starting point, Manferdini has developed new visual and spatial narratives that challenge perceptions of architectural environments through the use of decoration and ornamentation.

对于任何观察建筑的人，尤其是建筑师而言，似乎都对几何的力量有着无限的执着。这个几何学的领域，甚至说是最有效的领域就是网格：一个全面的，可控的系统。基于此系统，专业人士为我们构建出相当大一部分的建筑环境。现代美国城市见证了网格的重要性，它以解析和图解的方法表现出我们都能理解的，有秩序的全局组织规则。

该展览是一个系列展览的一部分，在此系列中，建筑和设计学院搜集了当代建筑师和设计师利用相关设施研究的建筑方案中的关键问题。曼费迪尼从密斯的简单网格化外观处理开始，在绘制了密斯建筑的正面图像以创建网格的数字绘图之后，她为芝加哥市设计了一个独特的天际线。曼费迪尼对网格的处理模糊了建筑文脉中时尚与图案之间的界限，并引入了一种与过去有着强烈联系的全新的当代景观。以历史为起点，曼费迪尼发展了新的视觉和空间叙事方式，通过使用装饰和纹饰来挑战人们对建筑环境的感知。

Building the Picture IV
Print on Powder Coated Aluminum
构建图片IV
在粉末涂层铝上打印
60 cm × 60 cm

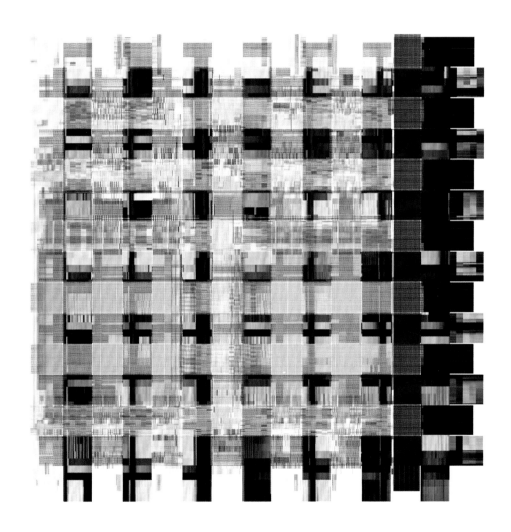

Building the Picture Ⅲ
Print on Powder Coated Aluminum
构建图片Ⅲ
在粉末涂层铝上打印
60 cm × 60 cm

Building the Picture V
Print on Powder Coated Aluminum
构建图片 V
在粉末涂层铝上打印
60 cm × 60 cm

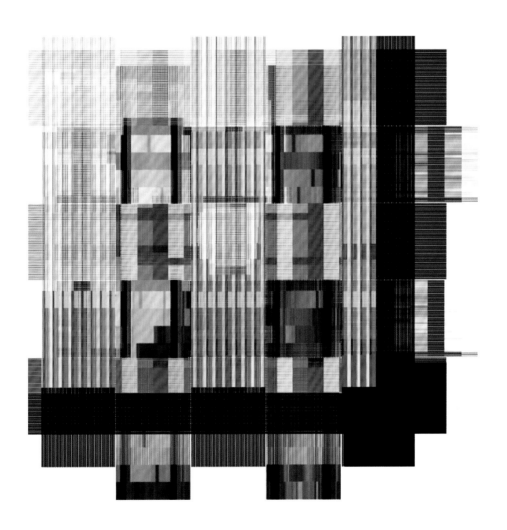

Grids and Typologies IV
Archival Ink Prints
网格和类型IV
档案墨水打印
48 cm x 33 cm

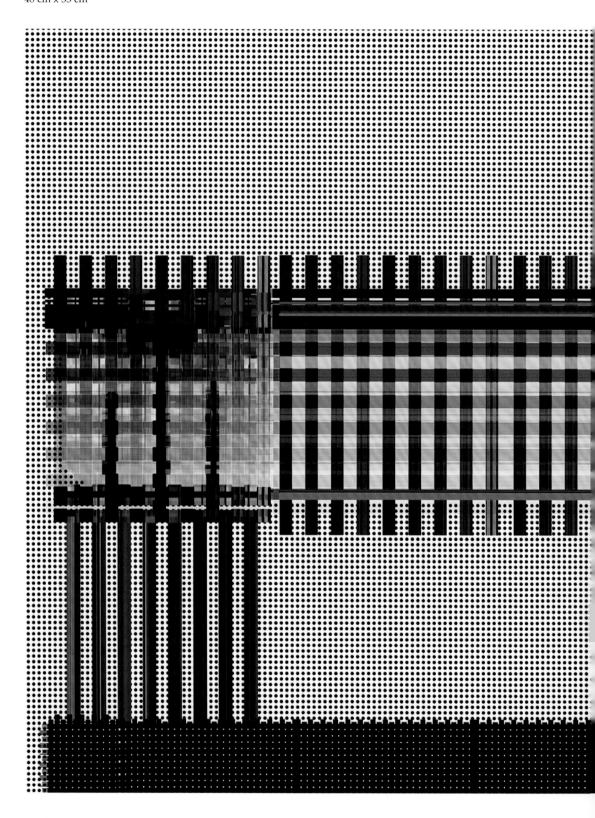

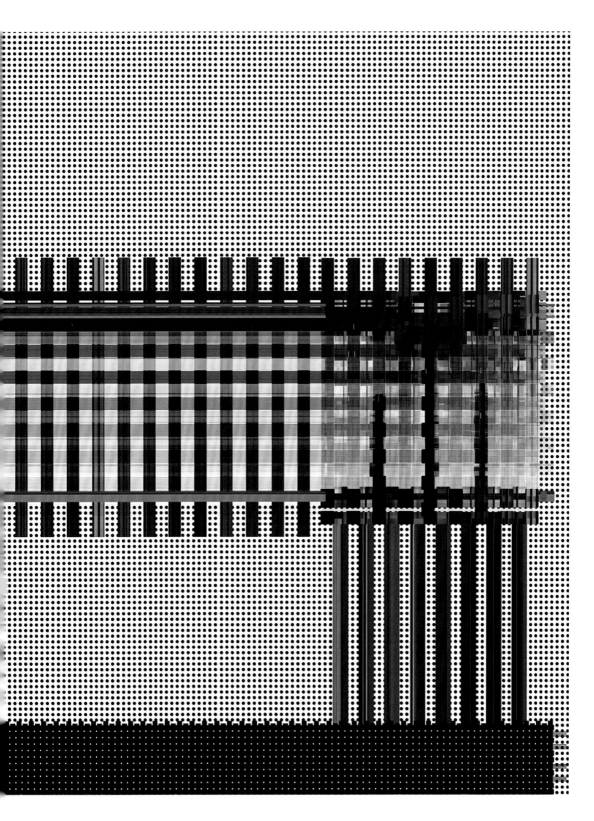

The capacity to produce a high-level of illusionism is today so advanced because of digital methods that it is less and less interesting to pursue. It's like hearing the same joke repeatedly: eventually it becomes monotonous. Architecture should not be too interesting. There is, in fact, a vastness of architecture that I refer to as "the practice of the week". Elena Manferdini is careful of not making this mistake. Her ability is in positioning her practice in between too interesting and not interesting at all. In terms of duration of attention that this practice generates, Manferdini's work requires a long span. The work holds back. Usually solid and well-trained architects are not interesting. In my opinion, her Miesian studies are quite amazing. Her best works are when she is perfectly willing to be bored. Boring is a perfect time for architecture.

Jeffrey Kipnis

如今，由于数字化的存在，产生高品质视觉体验的技术已经非常先进，因此人们对追求这种体验的兴趣也越来越低。就像重复听同一个笑话——其最终会变得单调乏味。建筑不应该太过有趣。事实上，绝大多数的建筑模式都被我称之为"周例法则"（原文为一周七天的建筑方案，意指常规而无聊的建筑行为）。埃琳娜·曼费迪尼小心规避着这个问题。她的才能是将她的方案控制在极度趣味性和完全无趣之间。就其方案产生的注意力持续时间而言，曼费迪尼的作品需要更长的时间跨度。通常来说，扎实、训练有素的建筑师大多无趣。在我看来，她对于密斯风格的研究非常出彩。她最好的作品是当她完全愿意展现无趣性的时候，无聊成了建筑学的最佳时机。

杰弗雷·基普尼斯

Grids and Typologies III
Archival Ink Prints
网格和类型III
档案墨水打印
33 cm × 48 cm

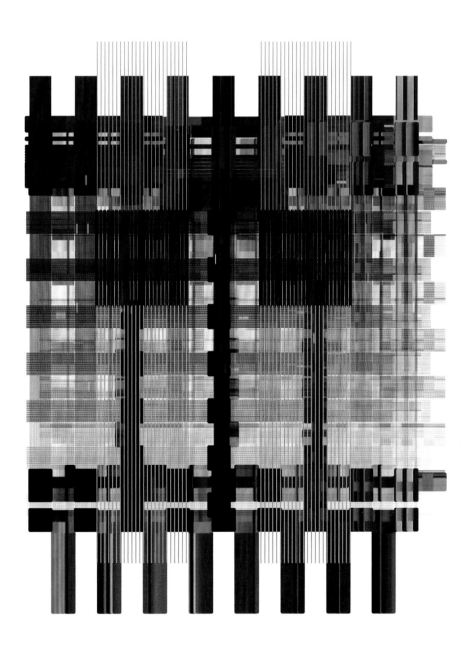

Building the Picture VI
Print on Powder Coated Aluminum
构建图片 VI
在粉末涂层铝上打印
60 cm × 60 cm

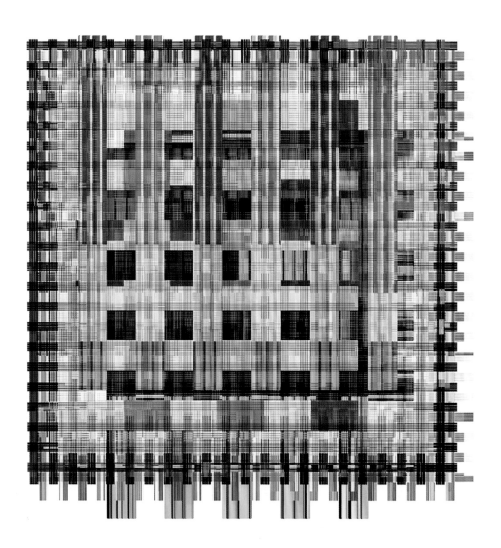

Building the Picture VIII
Print on Powder Coated Aluminum
构建图片VIII
在粉末涂层铝上打印
60 cm × 60 cm

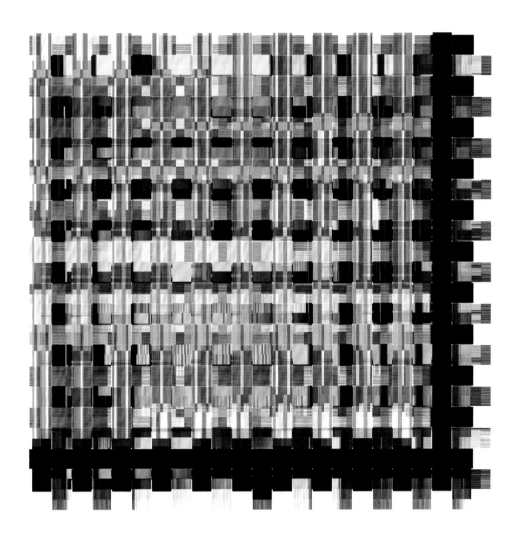

Elena Manferdini's work is surface-based but not cosmetic. It's not about producing a skin and infiltrating the skin. It's much more Miesian. I'm always struck by the proportions of the lines in the graphics. The graphics always have these neoplastic or tectonic qualities. It appears to be rendered quickly in mullions and in panels, as if it required no effort. The pictures always look like they could easily be done in ceramics or in wall painting, such as the ones of Tiepolo. If you look at the quality of the color and the rendering in the pictures, they have a mapped quality that keeps them from looking like a Rosenquist or a Jeff Koons piece.

Jeffrey Kipnis

埃琳娜·曼费迪尼的作品基于表面，但并不浮于表面。其作品不是创造表皮并渗透表皮的行为活动，更像是密斯的风格。我总被这些图形中的线条所震惊。这些图形总是具有一些肿瘤般的或构造般的特性。它得以在竖框和嵌板中快速呈现，就好像不耗费任何努力一般。这些作品很容易在陶瓷或墙壁上绘制，比如蒂耶波洛的几幅作品。如果你观察作品中颜色的质感和渲染效果，就会发现它们具有强烈的信息映射特质，使它们看起来不同于罗森奎斯特的或是杰夫·昆斯的作品。

杰弗雷·基普尼斯

Grids and Typologies Ⅱ
Archival Ink Prints
网格和类型 Ⅱ
档案墨水打印
33 cm × 48 cm

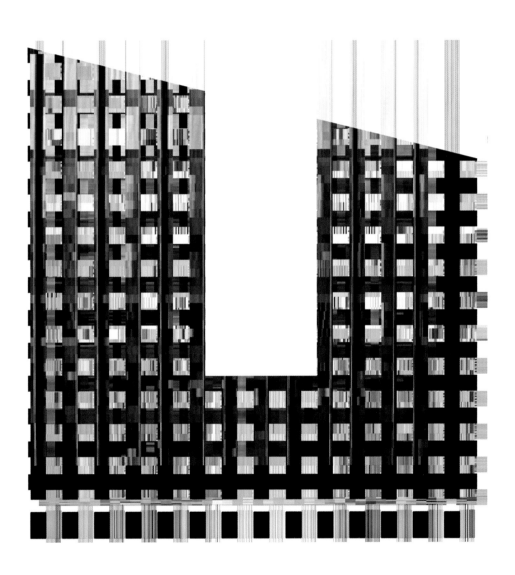

Grids and Typologies I
Archival Ink Prints
网格和类型I
档案墨水打印
48 cm × 33 cm

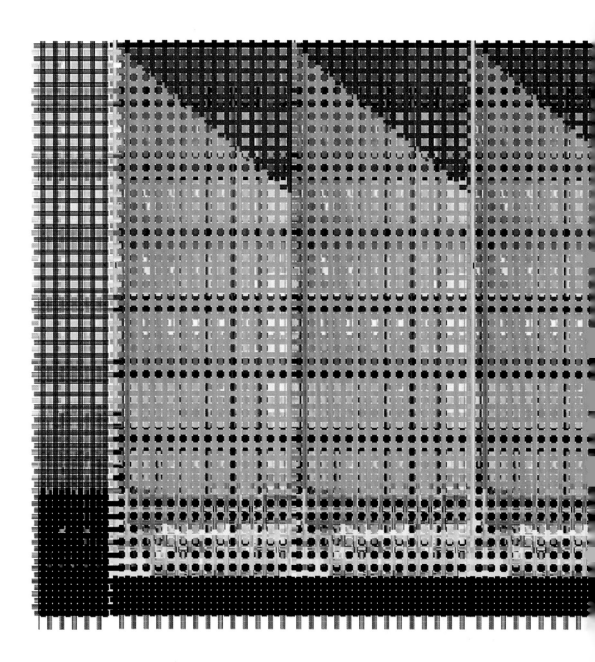

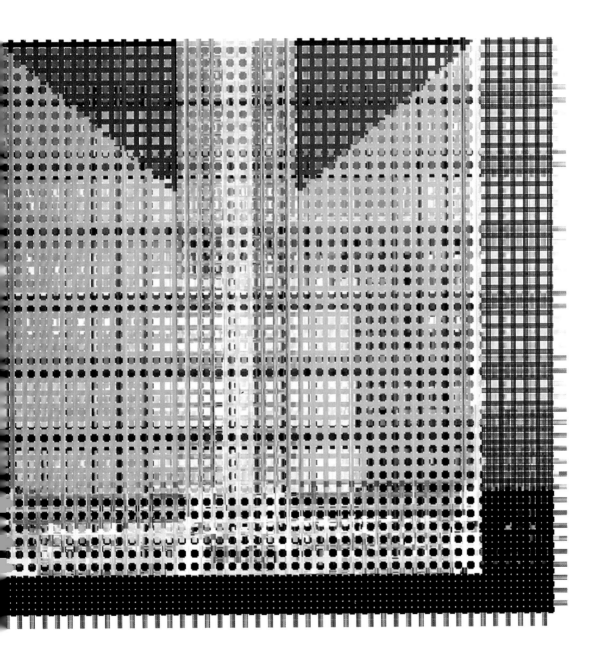

Building the Picture I
Print on Powder Coated Aluminum
构建图片 I
在粉末涂层铝上打印
60 cm × 122 cm

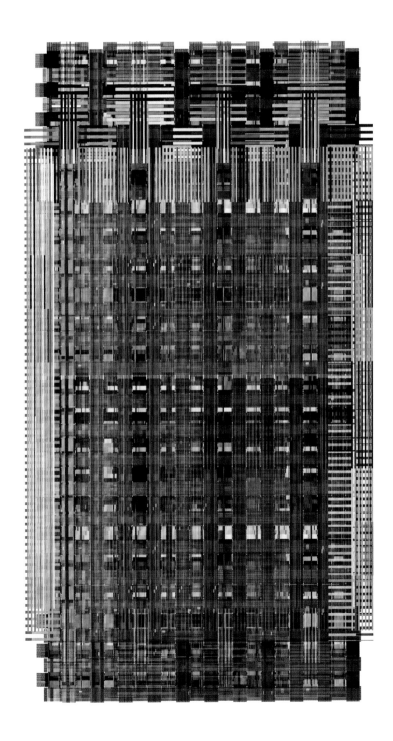

Building the Picture X
Print on Powder Coated Aluminum
构建图片 X
在粉末涂层铝上打印
60 cm × 60 cm

Building the Picture XII
Print on Powder Coated Aluminum
构建图片 XII
在粉末涂层铝上打印
60 cm × 60 cm

Building the Picture IX
 Print on Powder Coated Aluminum
构建图片IX
在粉末涂层铝上打印
60 cm × 60 cm

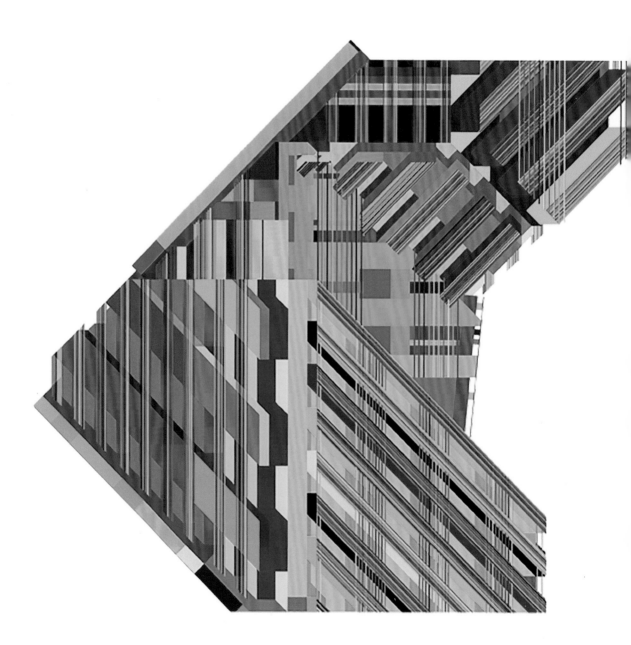

Building the Picture XI
Print on Powder Coated Aluminum
构建图片XI
在粉末涂层铝上打印
60 cm × 60 cm

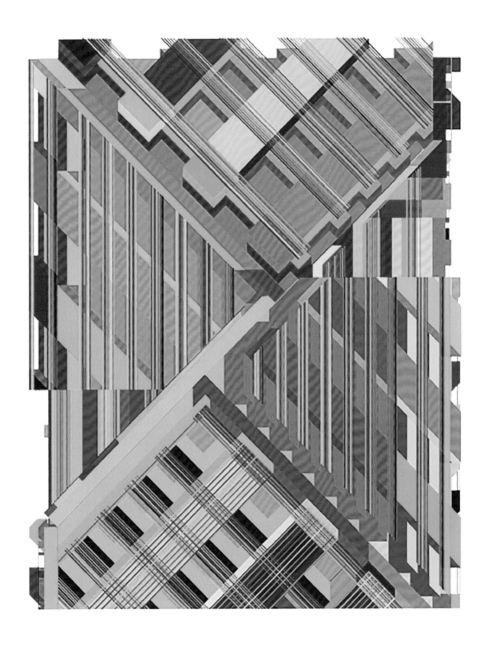

Building the Picture - Unfold I
Archival Ink Prints
物理模型的安排展开 I
档案墨水打印
33 cm × 48 cm

Building the Picture - Unfold II
Archival Ink Prints
物理模型的安排展开 II
档案墨水打印
33 cm × 48 cm

Building the Picture - Unfold III
Archival Ink Prints
物理模型的安排展开III
档案墨水打印
33 cm × 48 cm

Fictional Chicago Skyline 虚构的芝加哥天际线

Fictional Chicago Skyline虚构的芝加哥天际线

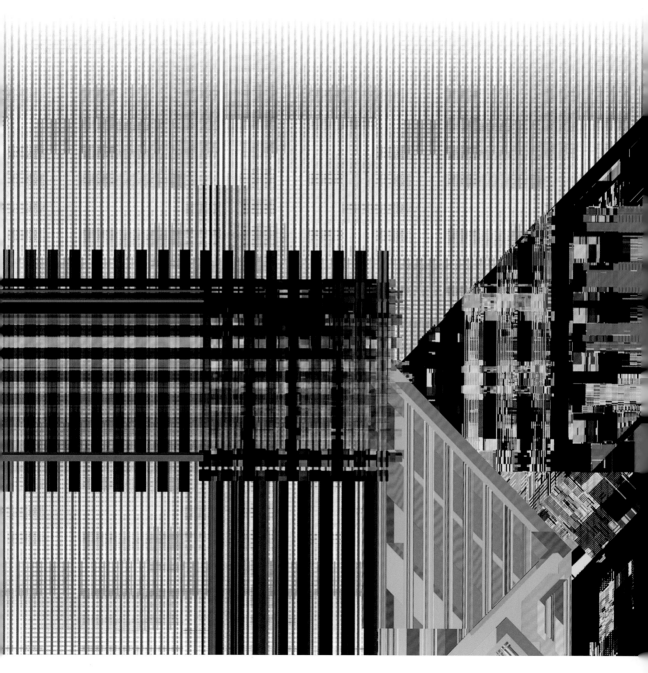

BUILDING PORTRAITS

INDUSTRY GALLERY

Date: 2016

Location: Los Angeles, California, USA

"Building Portraits" is a suite of elevation studies and physical models developed by Atelier Manferdini for a solo show at Industry Gallery, in Los Angeles, 2016. The 42 drawings were produced in rapid succession and they explore the potential of intricate scripted linework to depict building facades. The collection exists simultaneously as architectural research and as an autonomous artwork. These drawings can be understood as scaled down reproductions of buildings and, at the same time, as full scale printed artifacts. The collection plays with the graphic potentials of woven grids and scripted vector lines, while exploring the canonical relationships of shape vs form, ground vs figure, pattern vs coloration, orientation vs posture. The title of the suite, "Building Portraits", alludes to two distinct disciplines: the field of architectural drawings (building), and the one of fine artistic pictures (portraits). This body of work tries to claim a territory where these two attitudes find a common ground, where pixels and vectors get closer in scale of perception.

This research revolves around a specific relationship between digital scripted drawings and analog pictures of the architectural medium. This work makes a case that scripted drawings now have the ability to carry an enormous amount of data, and that therefore we are closing the gap between the analog mediums of architecture and drawings.

建筑肖像

工业画廊

时间：2016年

地点：美国，加利福尼亚州，洛杉矶

"建筑肖像"是曼费迪尼工作室为2016年洛杉矶工业画廊的个展而创作的一系列立面研究和实体模型作品。这42幅图画是在短时间内被连续绘制出来的，它们探索了用复杂的编程线条来描绘建筑立面的可能性。该系列作品既是建筑研究也是独立的艺术品，这些画作可被看作建筑物的缩小复制品，同时也可被看作全尺寸的艺术品。该系列发掘了编织网格和脚本化矢量线的视觉潜能，同时探索形状与形式、底线与图形、图案与着色、方向与姿势的典型关系。该系列的标题"建筑肖像"，暗指两个不同的领域：建筑绘图（建筑）领域和美术（肖像）领域。这项作品试图宣告一个全新领域的形成，在此领域两种理念找到了共同点，像素和向量在感知度上更加接近彼此。

这项研究围绕着数字脚本化绘图和建筑媒介的模拟图画之间的特定关系展开。该作品使得脚本化的绘图现在能够承载大量数据，因此我们正在缩小建筑的模拟媒介和绘图之间的差距。

Building Portraits V
SFMOMA Permanent Collection
Archival Ink Prints
建筑肖像 V
旧金山现代艺术博物馆永久收藏
档案墨水打印
33 cm × 48 cm

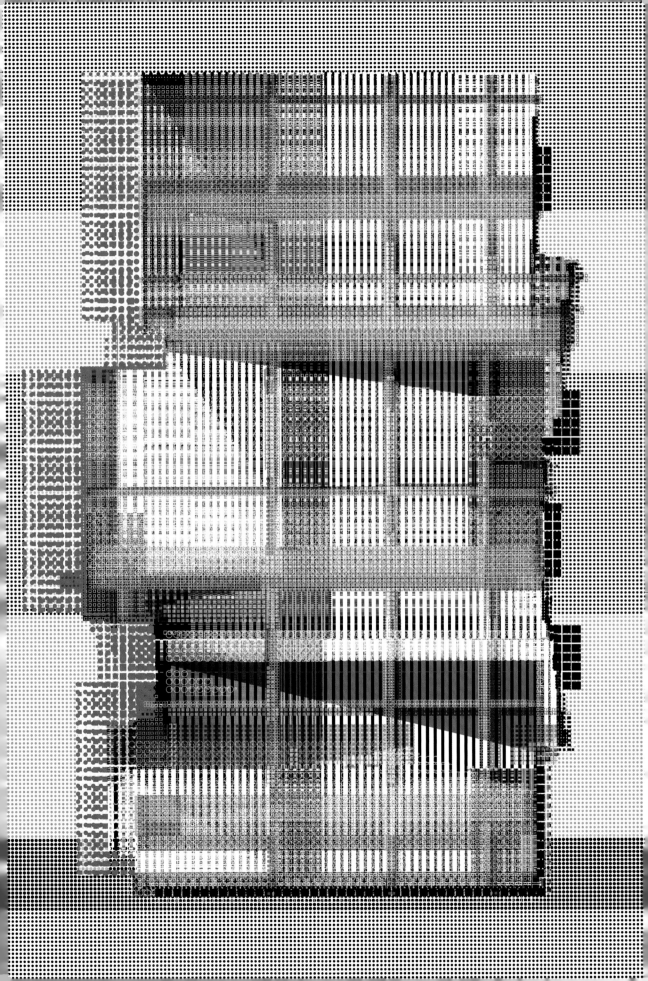

Digital tools have helped improve and potentiate the discipline and the practice of architecture, as well as facilitate the integration of active methodologies that every model requires 'coding.' Technological inputs can be found that fulfill the functionality of serving the practice and education, but very rarely, does the opportunity arise to find tools that can serve as robust inputs for the computer to become a critically essential tool that has drastically changed architecture culture.

Elena Manferdini's manifestation of her early 'coded' work shows a clear understanding this change. However, that specific digital culture (parametricism) has evolved in to 'pervasive order' or the wanted and unwanted dissemination of digital tools that have vastly transformed how we practice architecture and how we teach these tools which allow us to compare, create, and easily explore organized information. In that sense Elena Manferdini is a pioneer of 'pervasive' digital culture.

Gabriel Esquivel

数字工具有助于改善和增强建筑学科的理论和实践，同时促进每种模式所需"编码"的有效方法论的整合。技术输入可以产生促进实践和教育的功能，但我们很少能有机会找到为计算机提供可靠输入服务的工具，进而使其成为一个能够彻底改变建筑文化的至关重要的工具。

埃琳娜·曼费迪尼的早期"编码"作品证明她对这一变化有着清晰的认知。然而，这种特定的数字文化（参数化主义）已经演变成"普遍秩序"，或是对数字工具的必要和不必要的宣传，这些工具已经极大地改变了我们进行建筑实践、教学的方式，成为我们能够比较、创建和轻松地探索有序信息的工具。从这个意义上说，埃琳娜·曼费迪尼是"普及"数字文化的先驱。

加布里埃尔·埃斯奎维尔

Building Portraits VI
Archival Ink Prints
建筑肖像VI
档案墨水打印
33 cm × 48 cm

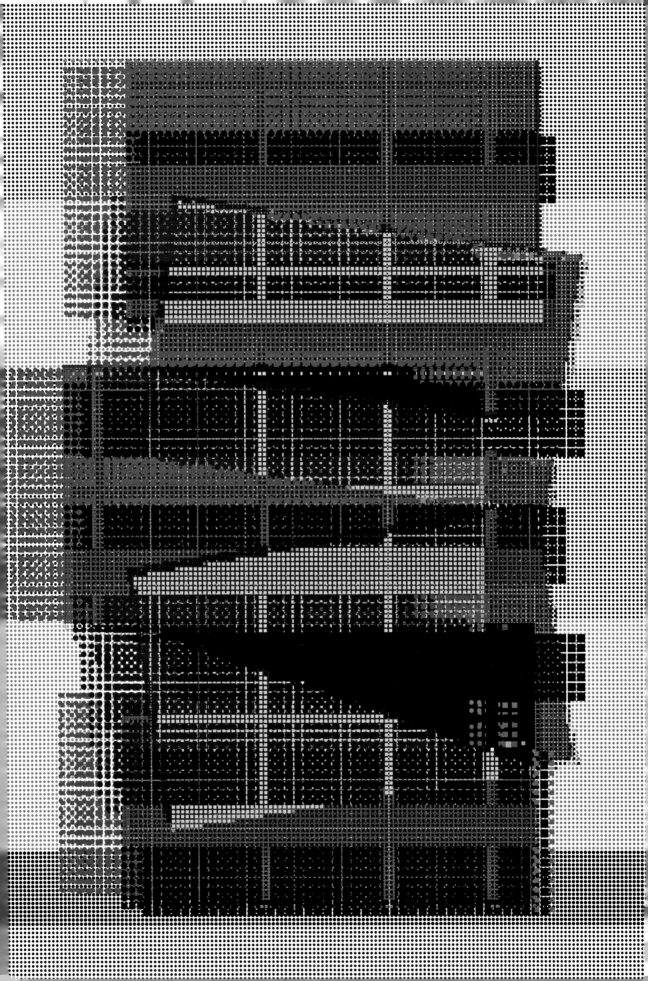

Building Portraits VII
Archival Ink Prints
建筑肖像VII
档案墨水打印
33 cm × 48 cm

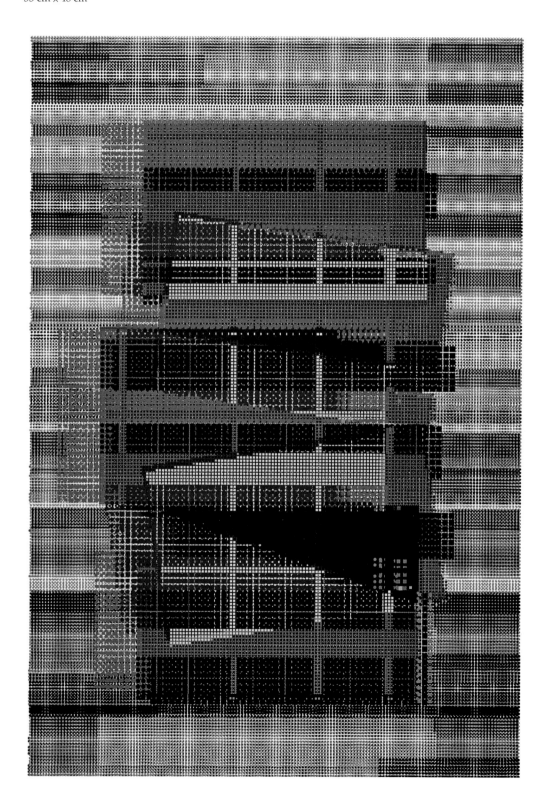

Building Portraits VIII
Archival Ink Prints
建筑肖像VIII
档案墨水打印
33 cm × 48 cm

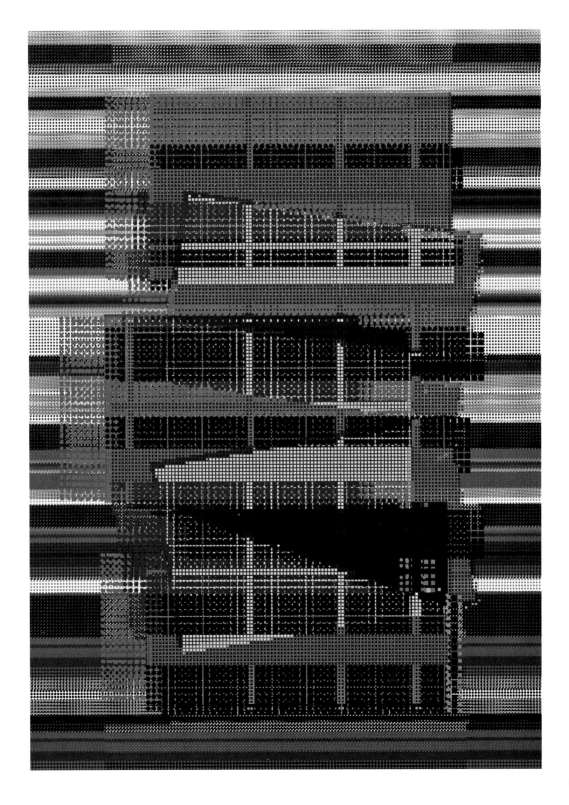

Fire Station - Ⅰ to Ⅳ
Archival Ink Prints
消防局—Ⅰ到Ⅳ
档案墨水打印
33 cm × 48 cm

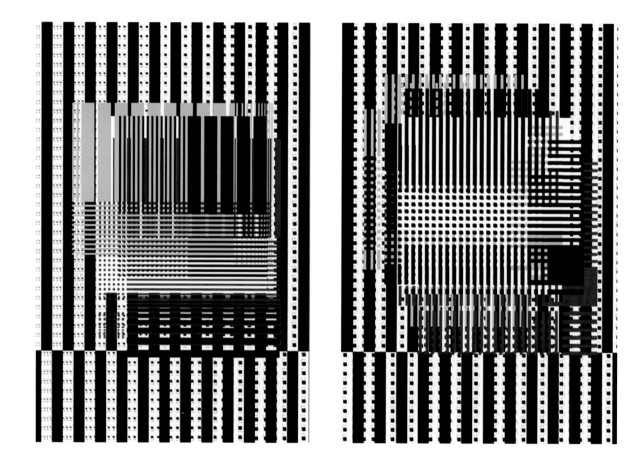

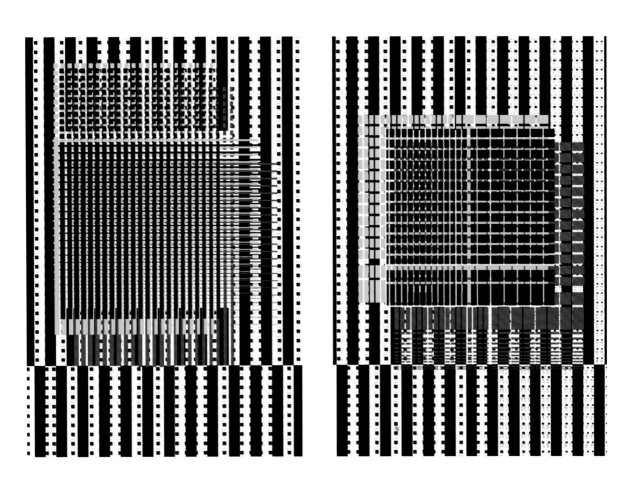

Postures and Ground Ⅴ
Archival Ink Prints
姿势和地面Ⅴ
档案墨水打印
33 cm × 48 cm

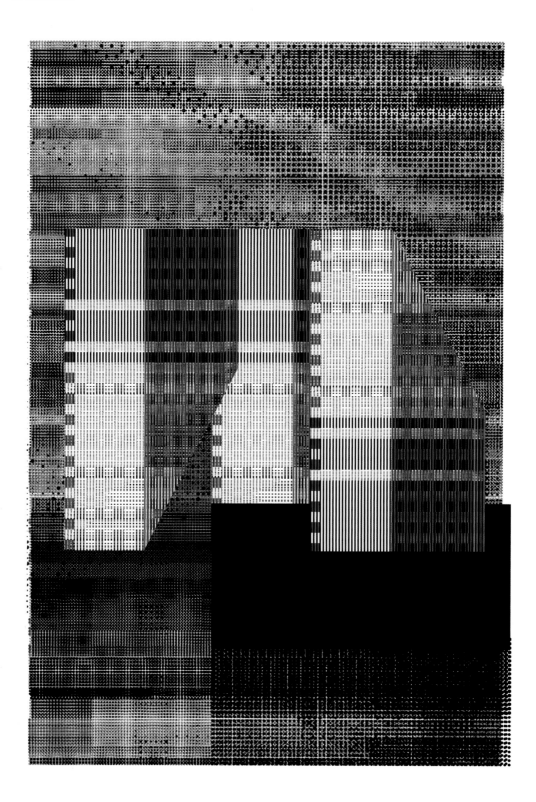

Postures and Ground VI
Archival Ink Prints
姿势和地面 VI
档案墨水打印
33 cm × 48 cm

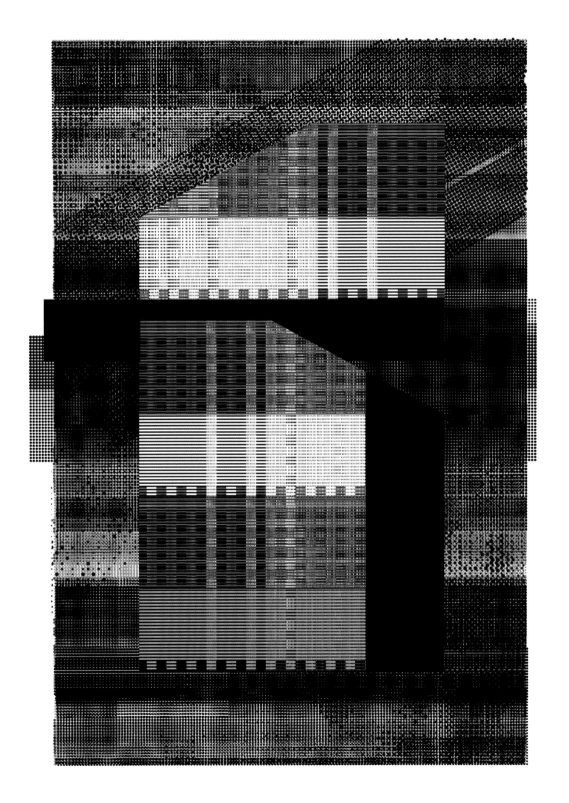

Elena Manferdini has a strong balance between the understanding and the respect of the discipline of architecture as a historical and precise discipline. Simultaneously, she has a strong interest in contemporary tools, in its way of thinking and producing. They may seem to be from different worlds, but she has the ability to blend them into one.

Hernan Diaz Alonso

埃琳娜·曼费迪尼在处理建筑学科的历史性和精确性之间拿捏得很到位。同时,她对现代化工具的思考和应用方面有着浓厚的兴趣。它们来自不同的领域,但她有能力将它们融合在一起。

埃尔南·迪亚兹·阿隆索

Postures and Ground I
Archival Ink Prints
姿势和地面 I
档案墨水打印
33 cm x 48 cm

Building Portraits IV
Archival Ink Prints
建筑肖像 IV
档案墨水打印
48 cm × 33 cm

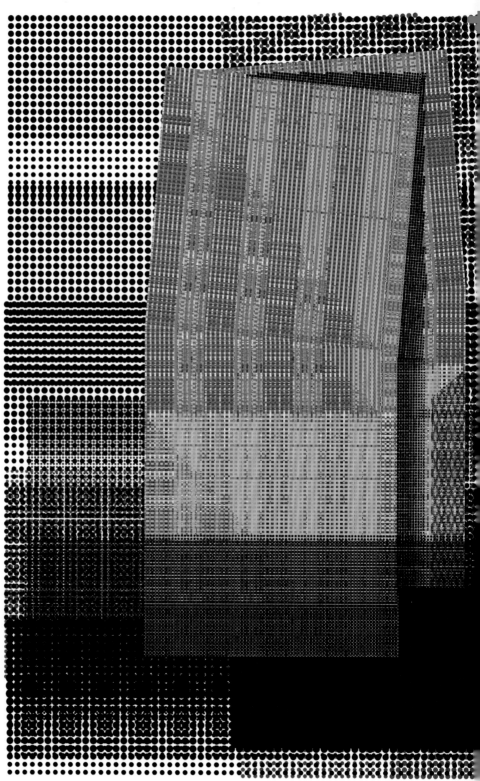

Digital tools have inarguably helped improve both the discipline and the practice of architecture. The wanted and unwanted dissemination of digital tools have vastly transformed how we practice architecture and how we teach these tools which allow us to compare, create, and easily explore organized information.

From the moment she started producing digitally fabricated projects, Elena Manferdini had a deep impact not only on the discipline but on the practice. Her work has gone through an evolution in terms of interests and focus. Every new direction she takes will become an influence to follow in every context.

Gabriel Esquivel

数字工具毫无疑问改善推动了建筑学科的理论和实践。数字工具的必要和不必要性的宣扬极大地改变了我们如何实践建筑的理念，以及我们如何教授这些使我们能够比较、创建和轻松地探索有序信息的工具。

从制作数字建造作品的那一刻起，埃琳娜·曼费迪尼对这门学科和其实践行为就产生了深刻的影响。她的作品在趣味性和关注点上都经历了蜕变，她所选择的每一个新方向都将会影响这个领域。

加布里埃尔·埃斯奎维尔

Shapes and Ground III
Archival Ink Prints
形状和地面III
档案墨水打印
33 cm x 48 cm

Shapes and Ground I
Archival Ink Prints
形状和地面 I
档案墨水打印
33 cm × 48 cm

Shapes and Ground Ⅱ
Archival Ink Prints
形状和地面 Ⅱ
档案墨水打印
33 cm × 48 cm

Building Portraits III
Archival Ink Prints
建筑肖像III
档案墨水打印
48 cm × 33 cm

INK ON MIRROR

INDUSTRY GALLERY

Date: 2016
Location: West Hollywood, Los Angeles, California, USA

镜子上的墨水

太平洋设计中心

时间：2016年
地点：美国，加利福尼亚州，洛杉矶，西好莱坞

"Ink on Mirror" was part of "Made in LA", the third iteration of the Hammer Museum's biennial exhibition which highlights the practices of artists working throughout Los Angeles and the surrounding areas.

For the show, Atelier Manferdini took 10 of the facade drawings developed for "Building Portraits" and made them three dimensional. The original, digitally scripted colored patterns were printed on reflective mirror and assembled in physical models. The use of mirror as support for the drawing came from the desire to express the role of the audience as pivotal in the discipline of architecture, and to invite the public's imagination into the picture plane of the drawing.

This is a silent protest against canonical elevations, which are structured so that one can read them standing vertically, from the top down, and from the center to the sides. Elevation drawings were, after all, the result of an architect standing still and looking at the world and imagining his/her contribution to it.

These mirror artifacts are not meant to be read vertically. The mirror creates a reflective pictorial surface that lets the world in, which acts in contrast to the heroic artist of the Renaissance who looks at nature, or the makings of Duchamp or Rauschenberg, who take their literal subject matter from the real world. Here the viewer and the building drawing enter into radical contact: the viewer inhabits the picture plane and becomes part of the drawing itself. Drawings are completed by our collective imagination.

"镜子上的墨水"是"洛杉矶制造"展览的一部分，该展览是哈默博物馆双年展的第三次迭代，旨在展示洛杉矶及周边地区的艺术家的作品。

在这次展览中，曼费迪尼工作室从为"建筑肖像"绘制的立面图中选取了10幅作品，将它们制成了三维形态。最初的数字编绘的彩色图案被印刷在镜面上，然后组装在实体模型中。使用镜子作为绘画的画布是希望表达公众是建筑学科的关键，并将公众的想象力引入画面中。

这是对传统立面的无声抗议，其结构使得人们可以垂直矗立，从上到下，从中心到两侧来解读。毕竟，传统的立面图是建筑师站着不动，看着世界，想象他/她对世界的贡献。

这些作品不应仅被垂直地解读。镜子创造了一个反射性的绘画表面来让世界融入其中，这与文艺复兴时期的那些着眼于自然的英雄主义艺术家形成鲜明对比，也与从现实世界中获取创作主题的杜尚和劳申伯格形成鲜明对比。在这里，观赏者和建筑图进行了彻底的交互：观看者处于在画作平面上，并成为画作本身的一部分——作品是靠我们集体想象而完成的。

Folds & Pleats II
Print on Acrylic
折叠和褶皱 II
在亚克力上打印
60 cm × 92 cm

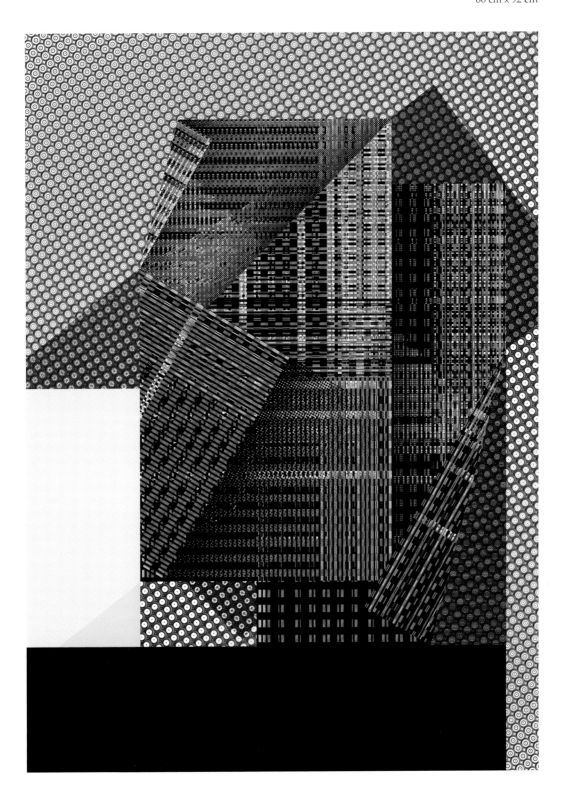

Folds & Pleats - III and I
Print on Acrylic
折叠和褶皱 - III和I
在亚克力上打印
60 cm × 92 cm

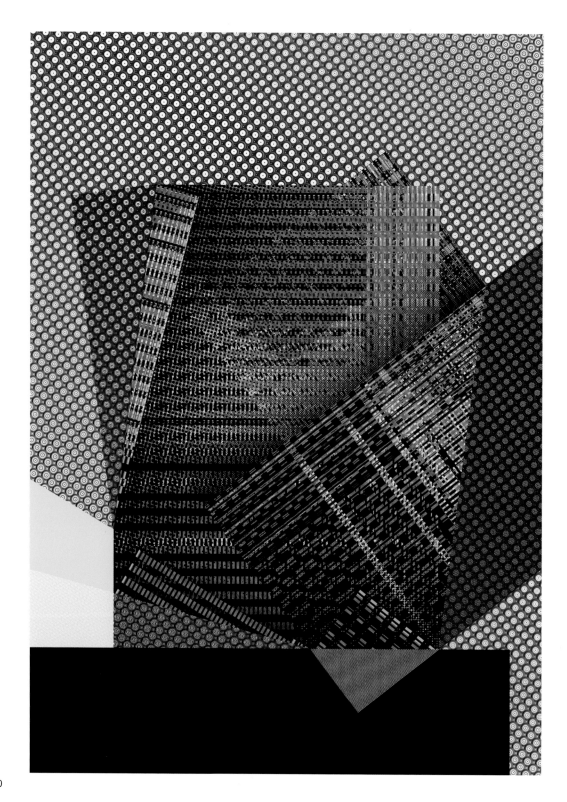

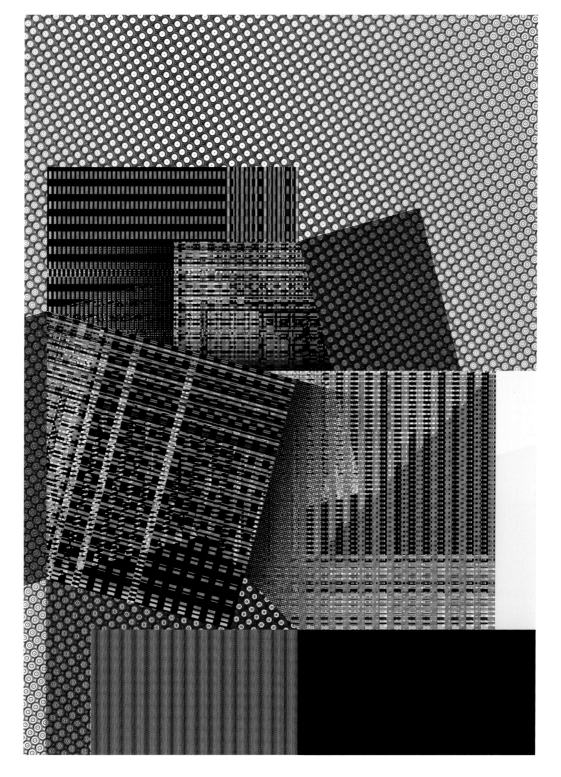

Elena Manferdini's shows are more interesting than any artist's show that's working with a lot of graphic and color. The details become the focal point which lessens the impact of the overall work. The details are where her work shines, and that's when the horizontal plane versus the vertical plane starts to become prevalent. It's honestly amazing.

Jeffrey Kipnis

埃琳娜·曼费迪尼的作品展比其他使用大量图形和颜色的艺术家的展览更有趣。作品中的细节成为减少整体作品影响力中的重要部分。细节是她的作品闪耀的地方，水平面和竖直面随处可见，真是太神奇了！

杰弗雷·基普尼斯

Building Portraits V - Model II
Wood Plinth and Acrylic Physical Model
建筑肖像V - 模型 II
木质底座和亚克力物理模型

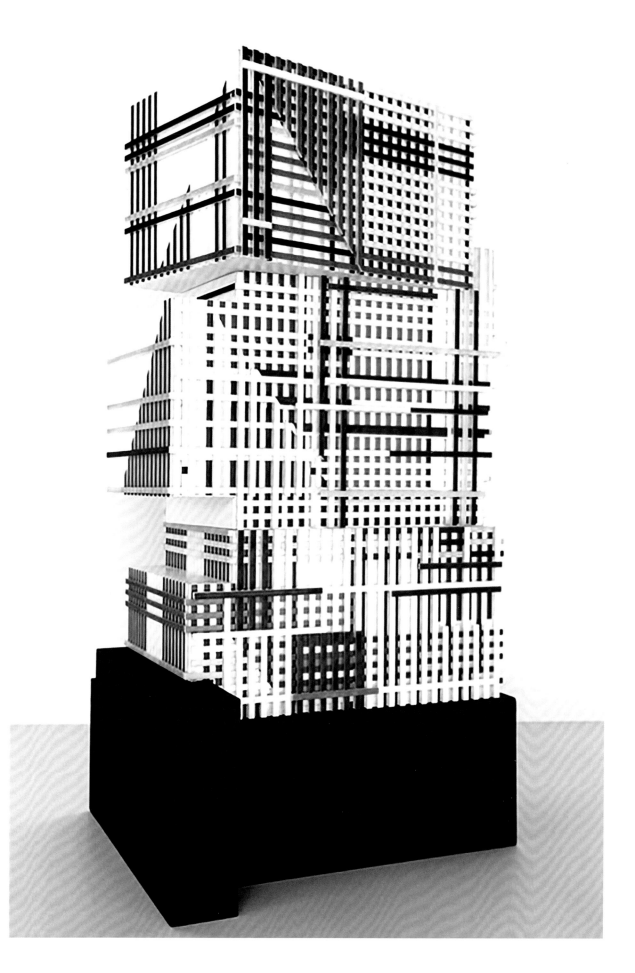

Postures and Ground Ⅱ
Archival Print
LACMA permanent collection
姿势和地面Ⅱ
档案打印
洛杉矶艺术博物馆永久馆藏
33 cm × 48 cm

Postures and Ground Ⅱ Model
Wood Plinth and Acrylic Physical Model
LACMA permanent collection
姿势和地面Ⅱ模型
木质底座和亚克力物理模型
洛杉矶艺术博物馆永久馆藏

Postures and Ground Ⅱ Unroll Drawing
Archival Ink Print
LACMA permanent collection
姿势和地面Ⅱ展开绘图
档案墨水打印
洛杉矶艺术博物馆永久馆藏
48 cm × 33 cm

Postures and Ground Ⅲ
Archival Print
LACMA permanent collection
姿势和地面Ⅲ
档案打印
洛杉矶艺术博物馆永久馆藏
33 cm × 48 cm

Postures and Ground III Model
Wood Plinth and Acrylic Physical Model
LACMA permanent collection
姿势和地面III模型
木质底座和亚克力物理模型
洛杉矶艺术博物馆永久馆藏

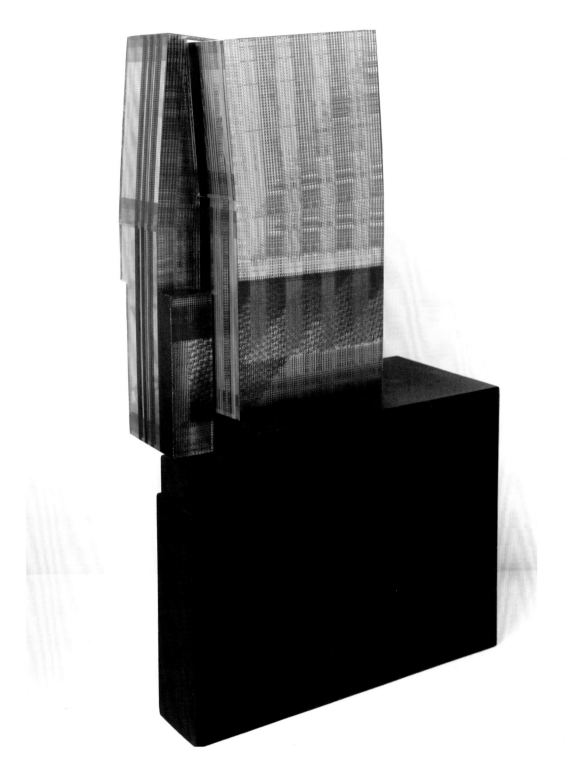

Postures and Ground Ⅲ Unroll Drawing
Archival Ink Print
LACMA permanent collection
姿势和地面Ⅲ展开绘图
档案墨水打印
洛杉矶艺术博物馆永久馆藏
48 cm × 33 cm

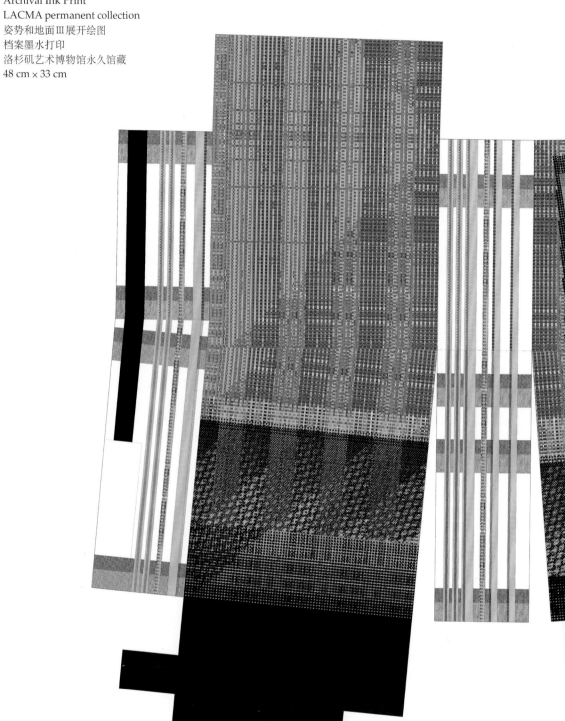

Forms and Ground – I to III
Archival Ink Prints
表格和地面 - I 至 III
档案墨水打印
33 cm × 48 cm

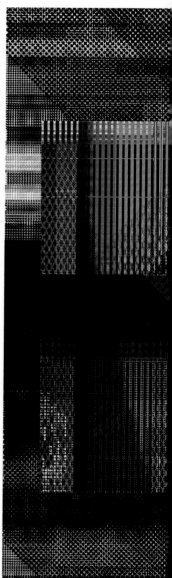

Forms and Ground I - III Models
Wood Plinth and Acrylic Physical Models
表格和地面 I - III 模型
木质底座和亚克力物理模型

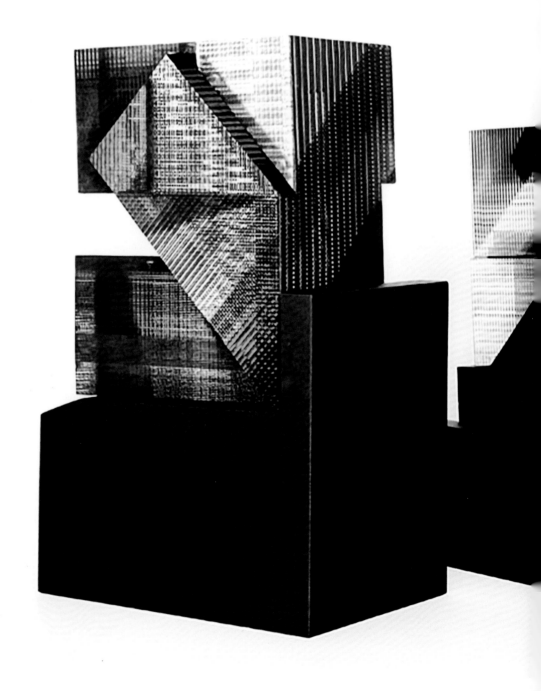

Forms and Ground Ⅱ Unroll Drawing
Archival Ink Print
表格和地面Ⅱ展开绘图
档案墨水打印
48 cm × 33 cm

Forms and Ground I Unroll Drawing
Archival Ink Print
表格和地面 I 展开绘图
档案墨水打印
48 cm × 33 cm

Postures and Ground IV
Archival Ink Prints
姿势和地面IV
档案墨水打印
33 cm x48 cm

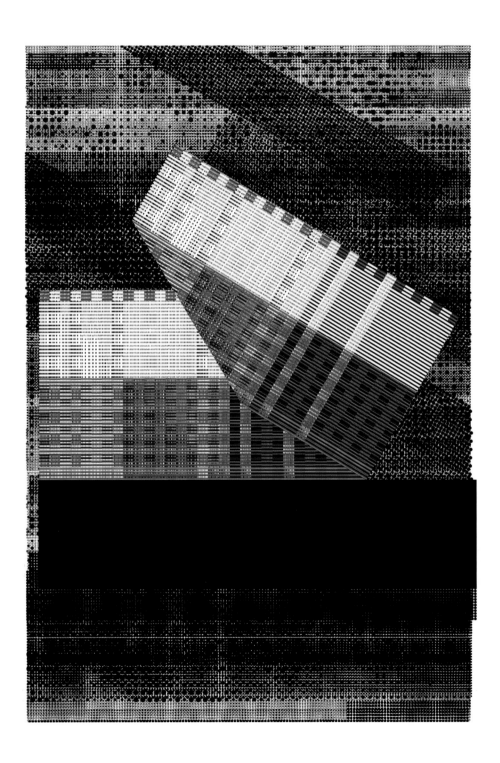

Postures and Ground Ⅳ - Model
Wood Plinth and Acrylic Physical Model
姿势和地面Ⅳ - 模型
木质底座和亚克力物理模型

Postures and Ground IV Unroll Drawing
Archival Ink Print
姿势和地面IV展开绘图
档案墨水打印
48 cm × 33 cm

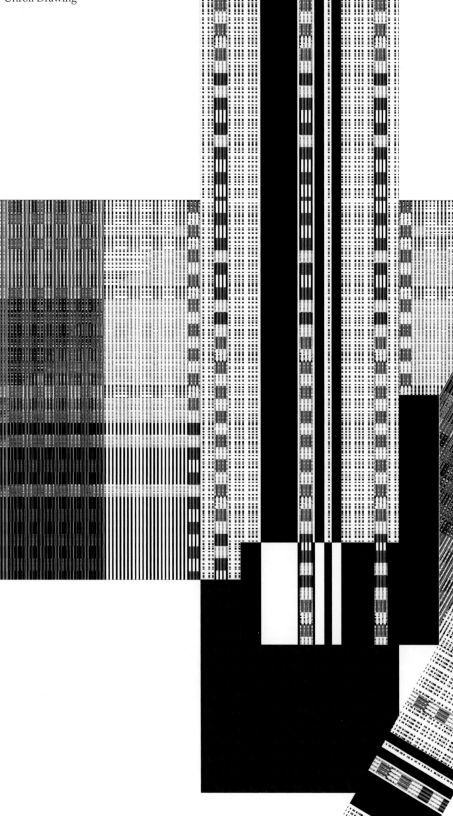

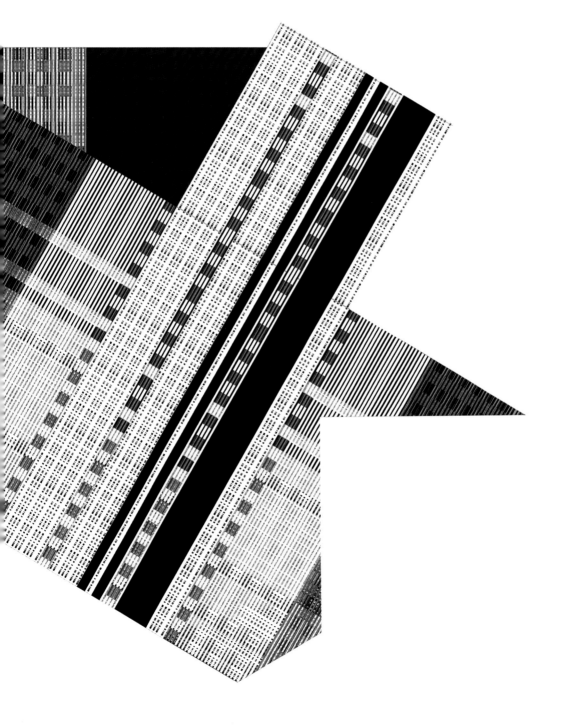

Postures and Ground V Unroll Drawing
Archival Ink Print
姿势和地面V展开绘图
档案墨水打印
48 cm x33 cm

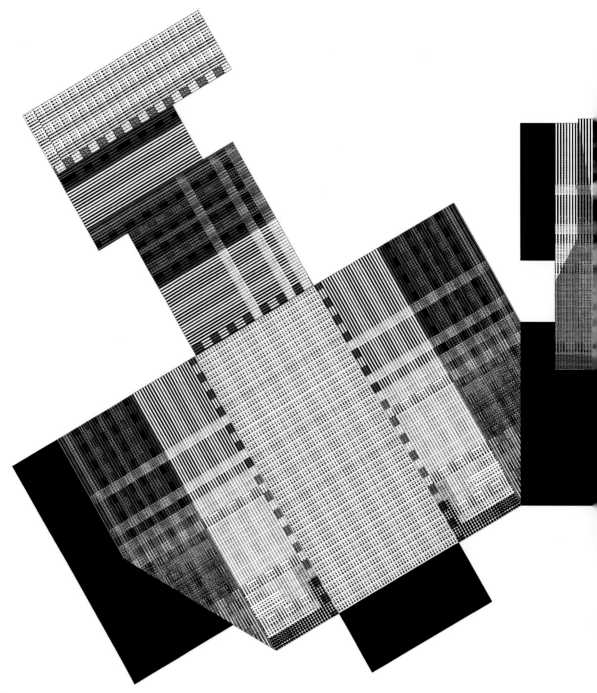

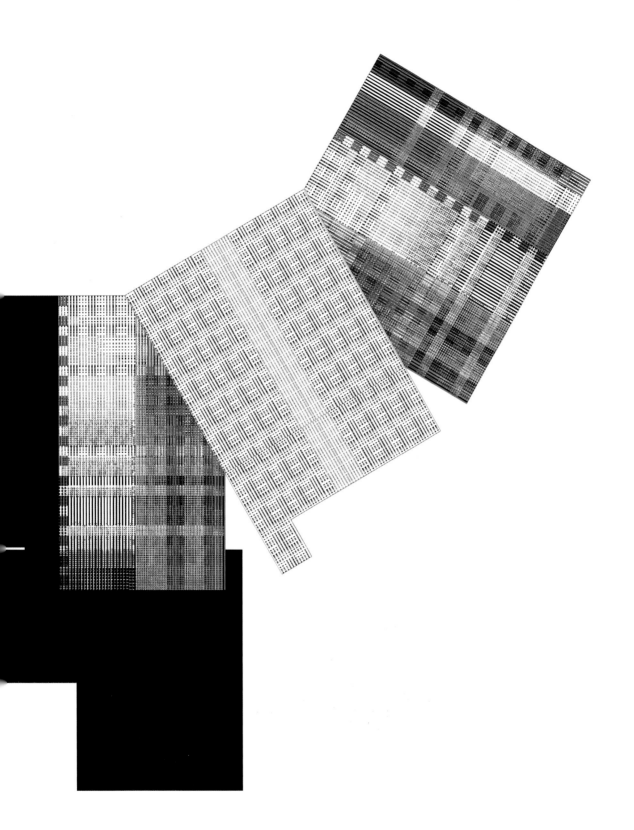

Postures and Ground Ⅳ - Ⅴ Models
Wood Plinth and Acrylic Physical Models
姿势和地面Ⅳ - Ⅴ模型
木质底座和亚克力物理模型

Forms and Ground Ⅲ Model
Wood Plinth and Acrylic Physical Model
表格和地面Ⅲ模型
木质底座和亚克力物理模型

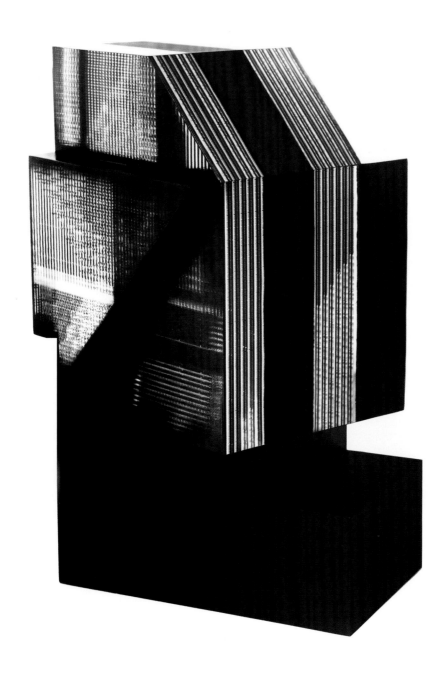

Forms and Ground Ⅲ Unroll Drawing
Archival Ink Print
表格和地面Ⅲ展开绘图
档案墨水打印
48 cm × 33 cm

There is literal space within a painting as well as space within the perspective of the painting. However, the real question is, "Where is the activation of the object emanating from and about?" The same thing occurs in architecture: there is a notion of a primitive plane, activated in our cultural, historic, civilized relationship to the art project of this discipline, with one being horizontal plane, and the other being the vertical plane. The painting is already telling the viewer how to look at it, and Elena Manferdini's work has a real power in claiming that it is a work of architecture. Le Corbusier was a painter, and almost became solely a painter. But he was never able to escape being an architect. The work of Elena Manferdini is the same, the horizontal planning throughout Manferdini's work can be felt. It's there, and she wants it to be there.

Jeffrey Kipnis

在这些作品中有真实的空间感，也给观察者留有想象的空间。然而，真正的问题是："客体本身的激发是从何产生，又关乎何事？"同样的事情也发生在建筑中：原始平面的概念，在我们与这种理念所处的文化、历史和文明关系中被激发，并映射在其水平面和垂直面上。这幅画作已经在告诉观众如何观赏它，而埃琳娜·曼费迪尼的作品在展示它作为一件建筑作品所具有的真正的力量。勒·柯布西耶是个画家，并且差点儿成为一个全职画家，但他永远也不能逃脱其建筑师的身份。埃琳娜·曼费迪尼的作品给人以同样的感觉，整个曼费迪尼作品中的横向布局都是可被感知的，它已经按照她希望的方式被呈现。

杰弗雷·基普尼斯

Forms and Ground V
Print on Powder Coated Steel, Gloss Coat
表格和地面V
在粉末涂层钢，光泽涂层上打印
60 cm × 92 cm

Forms and Ground VI
Print on Powder Coated Steel, Gloss Coat
表格和地面VI
在粉末涂层钢，光泽涂层上打印
60 cm × 92 cm

Forms and Ground IV
Print on Powder Coated Steel, Gloss Coat
表格和地面 IV
在粉末涂层钢，光泽涂层上打印
60 cm × 92 cm

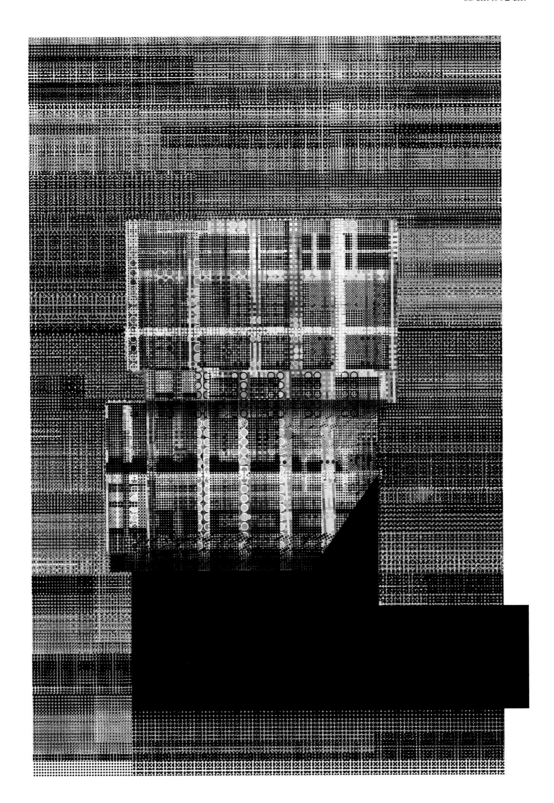

Forms and Ground IV Model
Wood Plinth and Acrylic Physical Model
表格和地面IV模型
木质底座和亚克力物理模型

Forms and Ground Ⅳ Unroll Drawing
Archival ink
表格和地面Ⅳ展开绘图
档案墨水
48 cm × 33 cm

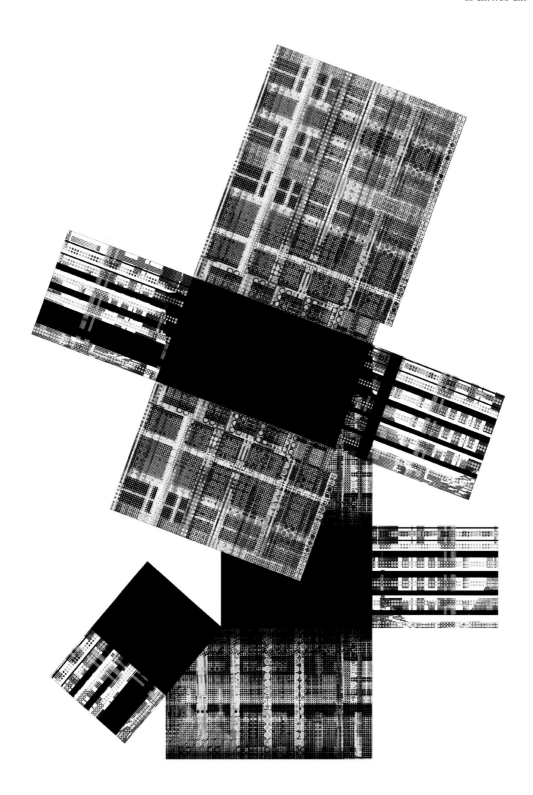

Forms and Ground VI Model
Wood Plinth and Acrylic Physical Model
表格和地面VI模型
木质底座和亚克力物理模型

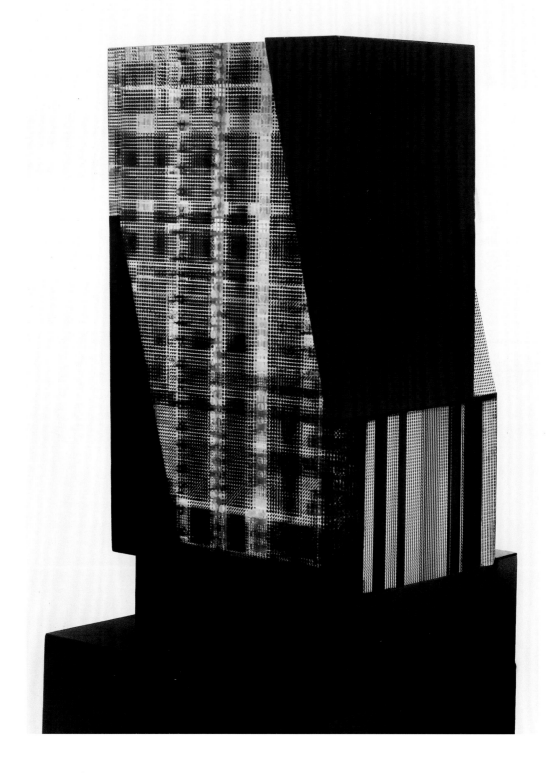

Forms and Ground VI Unroll Drawing
Archival ink
表格和地面VI展开绘图
档案墨水
33 cm × 48 cm

Suite of 9 Models
Wood Plinth and Acrylic Physical Model
9个套房模型
木质底座和亚克力物理模型

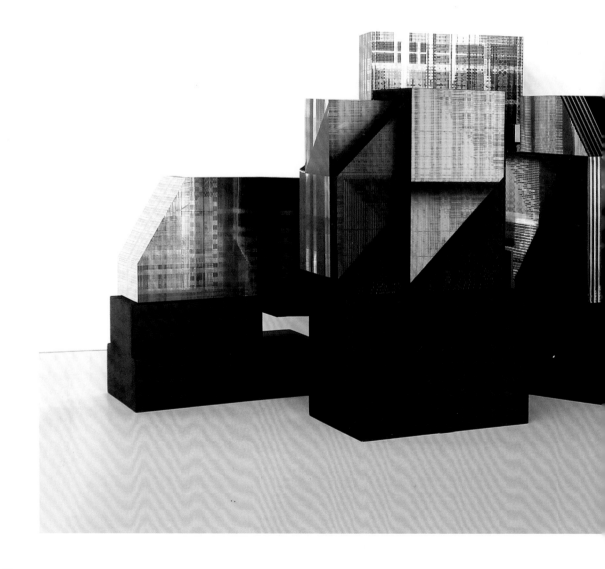

TWONESS

TOCO GALLERY

Date: 2016

Location: Los Angeles, California, USA

两元一体

图科画廊

时间：2016年

地点：美国，加利福尼亚州，洛杉矶

"Twoness" looks closely at the relationship between copy and original in architecture. In particular, the show builds upon the argument formulated by Derrida, that a copy produces the original as the original. The solo show exhibits 66 miniature printed drawings and argues that copy and original can coexist in what we will define as a state of twoness. The term twoness refers to a status of double consciousness and describes the feeling that you have more than one social identity, which makes it difficult to develop a sense of self.

To strengthen this position, one could argue that—as architects—we operate in a society of duplicates and copies, where originality is merely a romantic deception. Looking back at the practice of printing (silk screening, etching, stamping), copies have always been part of a process of disseminating ideas and creativity. The original is simply a tool to create copies. The printed copy was the intended output.

Nowadays, contemporary architectural practices have embraced a way of working that utilizes scripting chronicles, robotic fabrication, and digitized replicas even in the most personal part of the project: the sketch. Doubling is the new-normal that subverts the one-off and glorifies multiples. Mass production is our way of living. This series of drawings argues that it is time to focus theories on similarities rather than originality, and explores the relationship of replicas in the composition of the city.

"两元一体"密切关注建筑中的复制与原创的关系。具体来讲，该作品建立在德里达的论点基础上，即将一份副本产生的原创作为原稿。该作品展示了66幅微型印刷作品，并认为复制品和原作可以共存在被我们定义为两元一体的状态。两元一体一词指向一种双重意识的形态，它描述的是一个个体的多重社会身份感，这种多重感使得培养自我意识变得困难。

为强化这一立场，我们可以认为，作为建筑师，我们在一个仿制品和复制品的社会中工作，而在这个社会中，原创性仅仅是一场浪漫的骗局。回顾印刷的现实情况（丝网印刷、蚀刻、烫印），复制品一直是传播思想和创造力的重要部分，原件只是为了创建副本而存在的一种工具，印刷品才是我们真正想要的输出。

如今，现代建筑实践已经接受了使用脚本代码、机器人建造和数字化复制品的工作形态，甚至取代了在项目中最主观的部分——草图。内容的复制是一种新的常态，它颠覆了一次性创作，美化了多样性。大规模生产是我们的生活方式。这一系列的作品表明：是时候把理论集中在相似性而不是原创性上，并探讨复制品在城市构成中的关系。

26_Doppler_21,26_多普勒_21,18 cm × 13 cm

16_Doppler_13,16_多普勒_13,18 cm × 13 cm

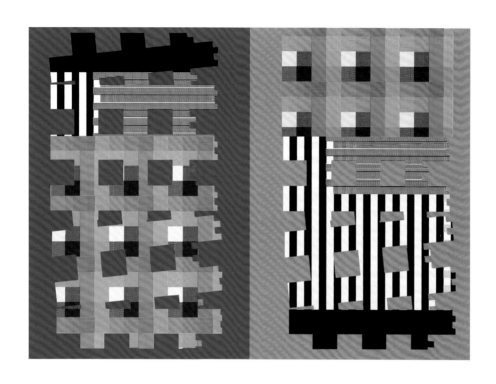

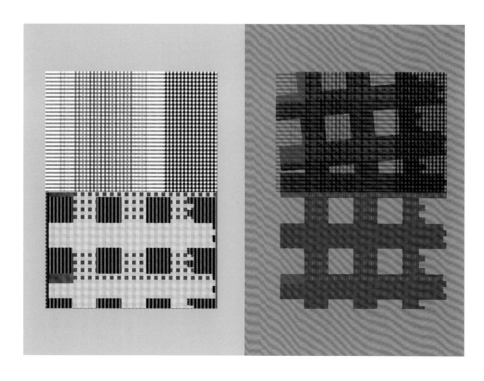

01_Doppler_01，01_多普勒_01，18 cm × 13 cm

02_Doppler_11，02_多普勒_11，18 cm × 13 cm

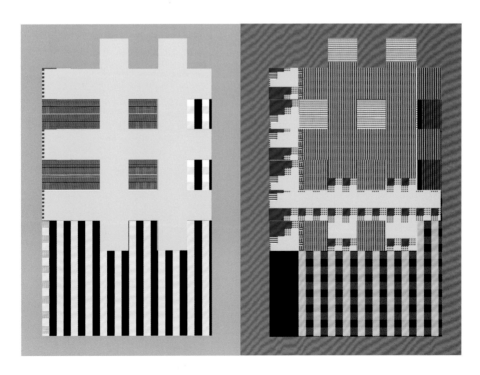

03_Doppler_02，03_多普勒_02，18 cm × 13 cm

04_Doppler_23，04_多普勒_23，18 cm × 13 cm

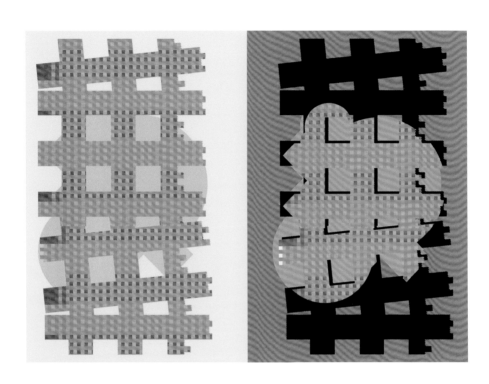

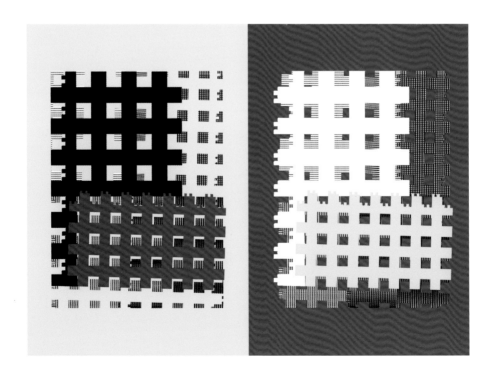

The artist's job is to answer questions we haven't asked yet.

Jeffrey Kipnis

艺术家的工作是回答我们尚未提出的问题。

杰弗雷·基普尼斯

07_Doppler_17，07_多普勒_17，18 cm × 13 cm

08_Doppler_05，08_多普勒_05，18 cm × 13 cm

Elena Manferdini does very interesting color combinations. She does triplets rather than duplets. That's why many aspects appear similar. That's why Rothko did 3,000 paintings. He did not have 3,000 ideas. Therefore, 90% of his paintings miss the mark. Occasionally, there is a piece that hits them all. He didn't know which paintings were going to be great and that's the direction that color is taking.

Nobody has ever taken color to the point in architecture where the color in architecture is the point of the architecture. The problem stems from how architecture is perceived in relation to color. "What's the role of color in architecture?" is a fantastically interesting question because, whether on a vertical surface or a horizontal surface, color helps to determine the phenomenology, the psychology, and the sociology in a horizontal plane that you're not even looking at.

Jeffrey Kipnis

埃琳娜·曼费迪尼进行了有趣的色彩组合。她做了很多三重色彩的搭配来取代二重搭配。这也可以解释为什么作品中的很多部分看来相似。亦如为什么罗斯科画了3000幅画，但他并没有3000个想法。因此，罗斯科90%的绘画都没有达到目标。但偶尔，会有一部作品打动所有人。他不知道哪些画会成为伟大的作品，这同样也是色彩使用的发展方向。

目前还没有人把色彩带到建筑领域中来声称建筑中的颜色是建筑的重点。这个问题源于建筑与色彩的关系。"色彩在建筑中的作用是什么"，这是一个非常有趣的问题，因为无论是在垂直的建筑表面还是水平的建筑面上，颜色都有助于用一个你不会注意到的方法，从根本上呈现建筑需要表现的那些现象学、心理学和社会学的内容。

杰弗雷·基普尼斯

15_Doppler_27,15_多普勒_27,18 cm × 13 cm

13_Doppler_20,13_多普勒_20,18 cm × 13 cm

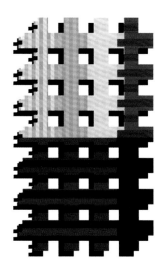
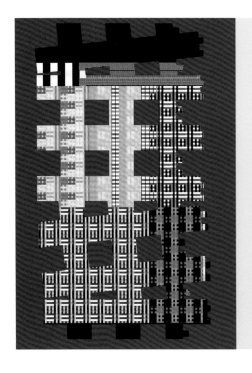
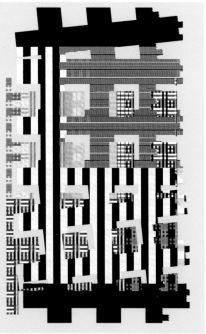

27_Doppler_25，27_多普勒_25，18 cm×13 cm

28_Doppler_29，28_多普勒_29，18 cm×13 cm

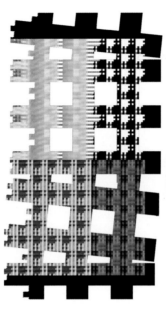
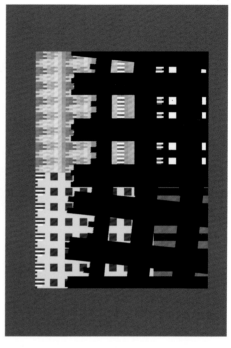

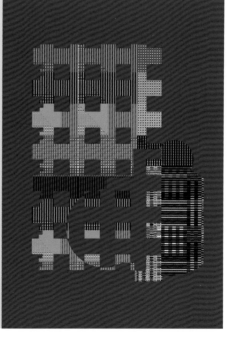

17_Doppler_14，17_多普勒_14，18 cm × 13 cm

20_Doppler_15，20_多普勒_15，18 cm × 13 cm

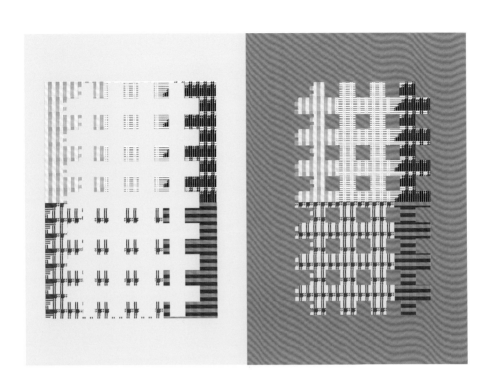

CHICAGO SKYLINE

CHICAGO ARCHITECTURE BIENNIAL

Date: 2017
Curators: Mark Lee and Sharon Johnston
Location: Chicago, Illinois, USA

芝加哥天际线

芝加哥建筑双年展

时间：2017年
策展人：马克·李和沙龙·约翰斯顿
地点：美国，伊利诺伊州，芝加哥

Exhibited as part of a group show of 140 practitioners, Atelier Manferdini's dimensional wallpaper builds from the earlier set of works, "Building Portraits". These portraits—ranging from small acrylic objects to fully immersive wallpapered environments—all originate from frontal photography of renowned Chicago buildings designed by Mies van der Rohe. Manferdini's interest in the portrait as a photographic mode lies in the insistence of the frontal view. This is a view that, in architectural representation, is most commonly referred to as the elevation: a measured drawing that is flat to the page and without the distortion of perspective. Using front on photographs of van der Rohe's Chicago towers, which were noted for their attention to the grid facade, allows Manferdini to bring the portrait photograph closer to the architectural elevation. We read the grid of the facade as flat to the page, while the tower's windows and supports become isomorphic with the printed page. Out of these portrait photographs, Manferdini's creations proliferate layers of Miesian grids, delaminating into a woven surface with manifold colors and interlaced textures; thus, the image becomes information to be worked on and transformed.

作为140名从业者参与的群体展览的一部分，曼费迪尼工作室的立体墙纸是根据早期的一组作品"建筑肖像"制作的。从小型亚克力材质到完全沉浸式的墙纸环境，这些肖像都源自密斯·凡·德·罗设计的芝加哥著名建筑的正面摄影。曼费迪尼对竖直作为一种摄影模式的兴趣潜藏在其对正面视角的坚持中。这种视角，在建筑表现中，常被称为立面：一种有尺度的绘图，与页面平齐，没有透视扭曲。使用密斯的芝加哥塔楼照片上的正面视角，也是因其以对网格立面的关注而闻名，这让曼费迪尼的作品更接近建筑外立面图。我们将立面的网格视为其被平放至所在面上，而塔楼的窗户和支架则与该打印页面同构。在这些竖直照片中，曼费迪尼的创作使密斯风格网格层出不穷，分层为具有多种颜色和交错纹理的线条编织的表面，至此，图像成为被处理和转换的重要信息。

Matrix of Folded Facades
Prints on Acrylic
矩阵和折叠立面
丙烯酸打印

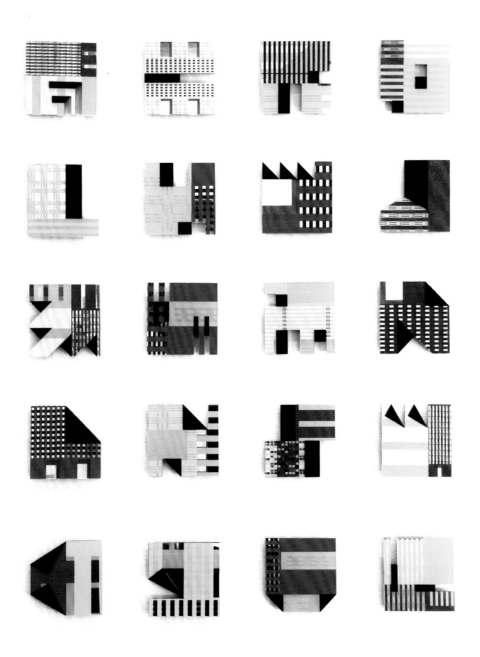

Elevations
Prints on Acrylic
立面
丙烯酸打印
122 cm × 244 cm

AYALA

FMB DEVELOPMENT

Date: 2019
Client: FMB Development
Location: Los Angeles, California, USA

阿亚拉

FMB地产开发公司

时间：2019年
客户：FMB地产开发公司
地点：美国加利福尼亚州，洛杉矶

One could argue that housing developments in Los Angeles are currently being dominated by passivity. By trying to appeal to everyone, they appeal to no one. Facades, even when buildings are privately owned, are important for the city at large because they are inevitably the background of our public imagination. The creation of "neutral" facades stifles this imagination by denying interaction and engagement with the social and cultural context. "Ayala" is a meditation on how the medium multi-family housing development can instead utilize its facade as a place where cultural relationships are resolved in an urban context. It proposes an alternative language for traditional facades based on vibrant color schemas and geometrical patterns which, along with Augmented Reality applications, aims to reinvigorate this important backdrop to the city. These 8 multi-family housing developments each offer their own pattern, adding uniqueness and specificity throughout West Hollywood. Where the traditional multi-family development would be flat, these patterns create both movement and depth, and instill an important sense of place for each.

有人可能会说，洛杉矶的住房开发现在更为被动，因为试图吸引每个人，反而对任何人都没有吸引力。即使建筑物是私人的，但立面对于整个城市也很重要，因为它们不可避免地成为公众想象的背景。"中性"立面的创造通过拒绝与社会和文化背景的互动和接触而扼杀了这种想象。"阿亚拉"是对中产阶级多户住宅的发展如何在城市环境中利用其外观来解决文化关系的思考。它基于鲜艳色彩模式和几何图案提出了一种传统外观的替代语言，与现实增强应用一起，旨在使这一重要景致重新焕发活力。这8个多户住宅开发项目各有特色，为西好莱坞增添了独特性和特殊性。传统的多户住宅的发展是平面化的，而这些方式带来了律动和深度，传递出一种强烈的位置感。

East Facade 东立面

North Facade北立面

West Facade 西立面

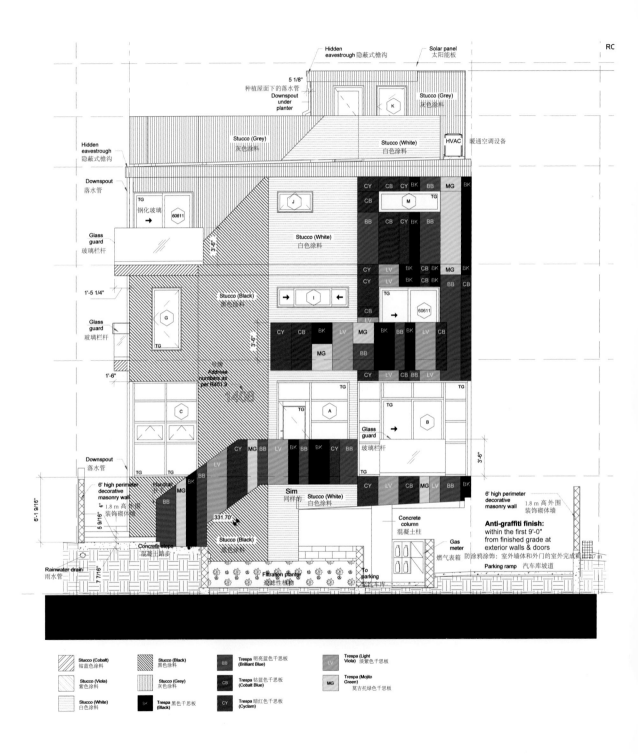

North Facade北立面

East Elevation 东立面

West Elevation 西立面

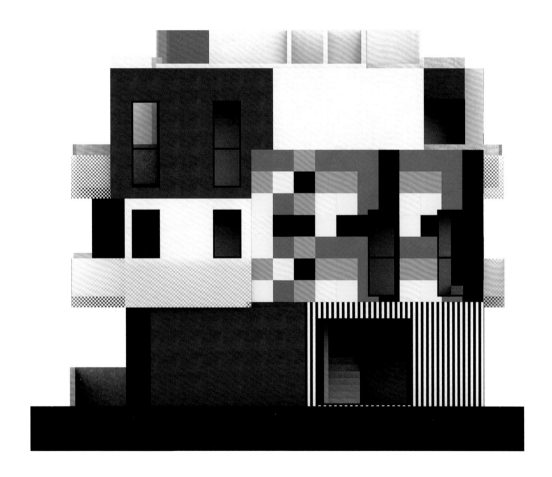

South Elevation南立面

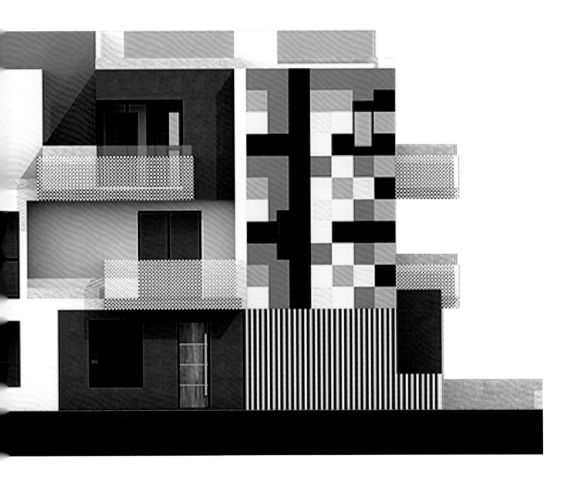

North and South Elevations 南北立面

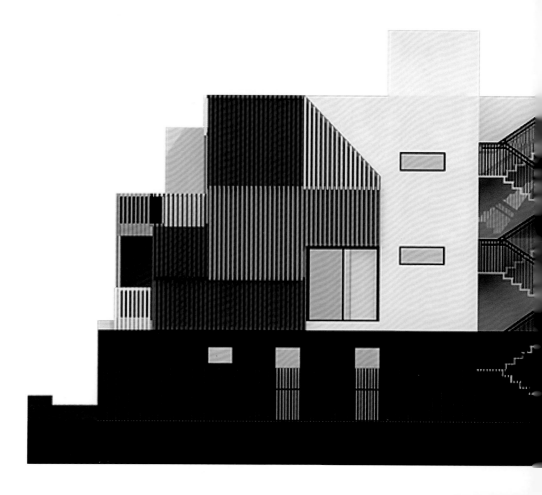

East Elevation 东立面

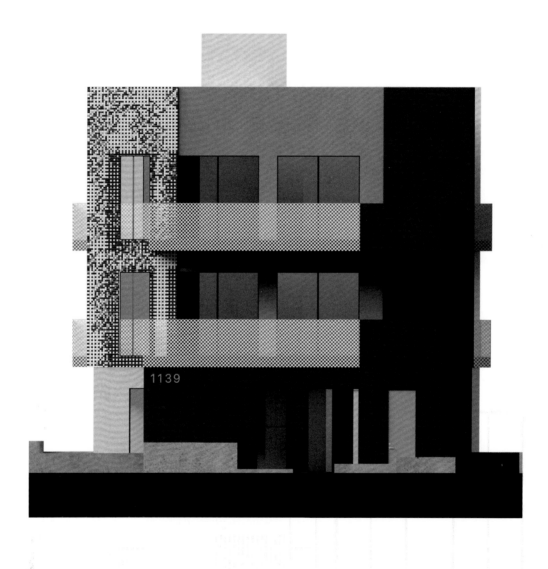

West Elevation西立面

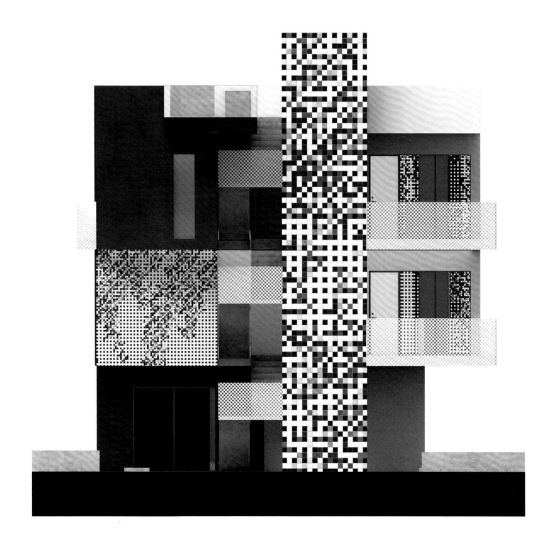

AT HUMAN SCALE

BMW

Date: 2015
Client: BMW
Location: Seoul, South Korea

Our love affair with automobiles has shaped our cities and our lives. Over a century on from one of the most revolutionary inventions of our time, one can see how the motor vehicle has determined the way in which modern cities have been planned, and also how traffic in time has become an important topic to address. In the future, there seems to be a need to create streets and spaces that make people feel they were planned for them. The human scale becomes the new choice—the only choice for a future urban experience. The installation picks up on this approach, exploring how the city changes when we put people at the center of our urban equations. The show strives to imagine what would happen if, in the future, cars, humans, and cities coexist to become one instead of existing in opposition to each other. The show is a vision of the architecture of the future city. Because in the future city humans will be at the center of the urban experience, in this exhibition the visitors feel larger than usual and cross the space like Gulliver in a new land.

人体尺度

宝马

日期：2015年
客户：宝马
地点：韩国，首尔

我们对汽车的热爱塑造了我们的城市和生活。一个多世纪以来，我们可以看到这个时代最具革命性的一项发明——汽车是如何决定现代城市的规划，以及我们怎样将省时交通作为重要议题来解决的。在未来，有必要创造街道和空间，让人们觉得这些空间是为他们而规划的。人体尺度成为新的议题，也是未来城市体验的唯一选择。整个建筑装置采用了这种理念，探索当我们把人放在城市这个方程的中心时，城市是如何变化的。该展览力求想象如果未来汽车、人类和城市共存而非相互对立时，会发生什么。这次展览是对未来城市建筑的展望，在未来的城市中，人类将成为城市体验的中心，因而在这次展览中，参观者感觉到他们在展览中的身形更加巨大，并像格列佛一样穿越在全新的空间中。

Digital Print IV
Archival Ink Print
数码打印IV
档案墨水打印
33 cm × 48 cm

Digital Print II
Archival Ink Print
数码打印 II
档案墨水打印
33 cm × 48 cm

Digital Print Ⅲ
Archival Ink Print
数码打印Ⅲ
档案墨水打印
33 cm × 48 cm

Installation安装

Installation安装

BLANK FACADE
HERMITAGE GARAGE

Date: 2018
Client: The Allen Morris Company
Location: Saint Petersburg, Florida, USA

空白立面
冬宫停车楼

时间：2018年
客户：艾伦·莫里斯公司
地点：美国，佛罗里达州，圣彼得堡

One of the initial "Building Portraits" drawings of Atelier Manferdini has been the point of departure for the design of a permanent art installation on a parking structure facade in Florida.

The 8 story garage is part of an 8 level, luxury rental apartment building, the Hermitage, located in downtown St. Petersburg. It hosts 348 living units. The artwork is conceived as a new façade applied to the core volume of the garage. Interestingly, parking facades are designed only "to a necessary degree": their appearance is often straightforward, earnest, and most important of all, non-exclusive. They are underestimated as anonymous, and this is one of the reasons their aesthetics remain independent from any stylistic movement. Nevertheless, they also own a strong identity that inadvertently "imposes" itself (Walter Gropius calls this "the unintentional beauty of industrial buildings"). These facades display a specific "brut" aesthetic that exists amongst architecture in the absence of routine human interaction. This aura develops over time. As this happens, such buildings emerge out of scale volumes.

The art façade, with its colorful fenestrations, brings a sense of human habitation to the existing garage's blank façade, which would otherwise be mute. The artwork introduces a new and contemporary sensibility to the place between graphic, deep hues and shallow depth.

曼费迪尼工作室最初的"建筑肖像"画作之一是这个佛罗里达州某停车楼建筑立面的永久性艺术装置设计的出发点。

这座8层的停车楼位于圣彼得堡市中心的一座8层豪华公寓楼"冬宫"中，公寓拥有348个居住单元。这件艺术品被设想成全新立面被应用于停车楼的核心主体中。有趣的是，通常，停车楼的立面被设计为"只基于其必要部分"——它们的外观是直白、严肃且非独特的。它们被平庸化，这也是它们的美学价值独立于其他风格化运动的原因之一。然而，它们也拥有一个不经意间"强化"自己的强大身份（沃尔特·格罗皮乌斯称之为"工业建筑的无意之美"）。这些立面在所有缺乏与人互动的建筑中展示了一种特殊的"粗犷"美学。这种观感随着时间的推移而发展，当这种情况发生时，这些建筑物就会在各处"生长"出来。

这种艺术立面，以其多彩的开窗给现有停车楼空白立面带来了一种人类居住的感觉，若外观没有这些改变，则会是毫无生气的。这件艺术品为介于图形、深色调和浅深度之间的部分引入了一种新的现代感。

South Elevation 南立面

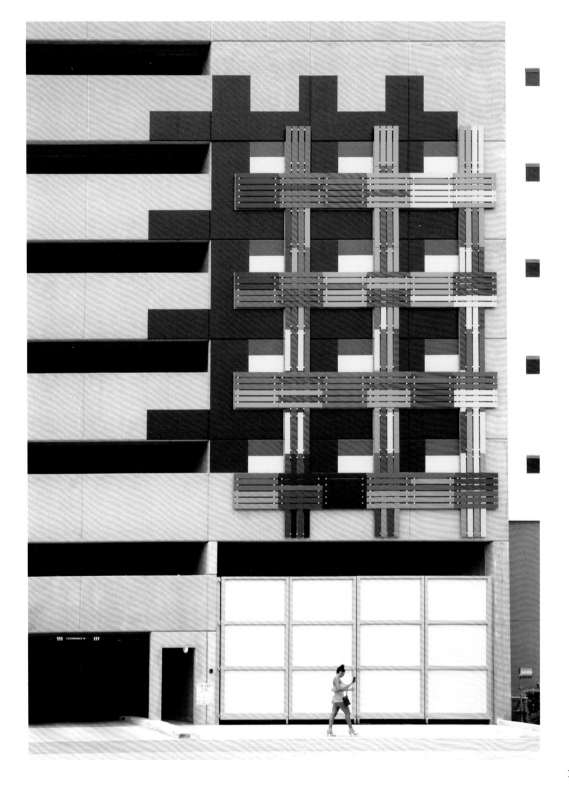

Facade Detail 立面细节

ALEXANDER MONTESSORI SCHOOL
FACADE SHADING PANELS

亚历山大蒙特梭利学校
立面遮阳板

Date: 2016
Client: James and Joyce McGhee
Location: Ludlam Road Campus, Miami, Florida, USA

时间：2016年
客户：詹姆斯和乔伊斯·麦吉
地点：美国，佛罗里达州，迈阿密拉德姆路校区

Alexander Montessori School is one of the largest Montessori institutions in the country, serving children aged 18 months to 12 years. The Montessori Method and philosophy of education was formulated in the early 1900's by Dr. Maria Montessori, an Italian engineer, psychologist, physician, and anthropologist. She believed that the purpose of education was not the imparting of knowledge by the teacher, but the development of intellectual skills so that the children could acquire the knowledge by their own effort, and thus making the knowledge truly their own.

In this unique type of school, the children work within a structured environment, filled with motivations for learning, where they are free to choose their work and to learn to make decisions. The development of a positive attitude in the child is the fundamental objective.

The design of the shading panels takes the hibiscus flower as a point of departure from the expected. The natural motif has been chosen by the school to complement its unique, hands-on personal learning experience of the children from nature. Floral figures are abstracted into pop cut-outs, and formally arranged to tie together the two buildings' façades and the connective corridor in between.

亚历山大蒙特梭利学校是美国最大的蒙特梭利学校之一，为18个月以上12岁以下的儿童提供服务。蒙特梭利教育方法和理念是在20世纪初由意大利工程师、心理学家、医生和人类学家玛丽亚·蒙特梭利博士制定的。她认为教育的目的不是教师传授知识，而是培养孩子的智力技能，使他们能够通过自己的努力获得知识，从而使知识成为自己的一部分。

在这个独特的学校里，孩子们处在一个有条理的环境中，充满了学习的动力，在这里他们可以自由地选择自己的课程，学习如何做出选择，培养孩子的积极态度是根本目标。

遮光板的设计以芙蓉花为出发点。学校选择了自然主题，以补充其独特的、亲身实践的个人学习体验。花卉图案被抽象表达成弹出式，并有规律地进行排列，将两栋建筑的外观和中间的连通走廊整合在一起。

Facade Shading Panels 立面遮阳板

141

Night View 夜景

Interior View 内部视图

CABINET OF WONDERS

LA PEER HOTEL

神奇陈列柜

拉佩尔酒店

Date: 2017
Client: Kimpton
Location: West Hollywood, California, USA

日期：2017年
客户：金普顿
地点：美国，加利福尼亚州，西好莱坞

Atelier Manferdini's proposal for the La Peer Entry Gate is a contemporary version of the renaissance "Cabinet of Wonders." These encyclopedic collections of curious items from home and abroad were filled with preserved animals, minerals, optical contraptions, as well as many other interesting man-made objects and specimens from exotic locations.

La Peer Entry Gate plays with the notion that the door is a place of discovery and wonder—where the gate becomes an artifact through which viewers can imagine multiple worlds hidden behind. As guests travel from near and far to stay at La Peer Hotel, the gate plays on the idea that every individual hosts a collection of memories, experiences, artifacts, and souvenirs. "Cabinets of Wonders" are—after all—the narrative of our human adventures, whether fictional or real.

La Peer Hotel Entry Gate contains a collection of bird figures that connect the viewer to other memories or thoughts. This connection to alternate realities is achieved via QR codes; thus using technology to reimagine the emotions evoked from these traditional cabinets. Unique QR codes are printed onto the birds, with each being programmed to contain links to various websites. When a user scans the code, they are transported to an alternate reality: a reality filled with video, sounds, or experiences.

曼费迪尼工作室的拉佩尔酒店大门设计方案是文艺复兴时期"神奇陈列柜"的一个当代版本。在这些来自国内外的百科全书式收藏品中充满了各种珍藏的动物（标本）、矿物、光学装置以及许多其他有趣的人造物品和异域标本。

拉佩尔酒店大门的设计概念认为门是一个探索和漫游的地方，门变成了一件艺术品，通过它，观众可以想象隐藏在门背后的多个世界。当客人从各处前往拉佩尔酒店住宿时，酒店大门传达了这样一个理念：每个人都拥有一系列的记忆、经历，他们的生活中留存了很多手工艺品和纪念品。"神奇陈列柜"无论虚构或是真实，都是人类冒险的故事。

拉佩尔酒店入口大门的设计包含一系列鸟的图案，将观众和他们的记忆联系起来，这种联系是通过二维码实现的。因此，可以使用科技来重新幻想从这些传统陈列柜中唤起的情感。这些独特的二维码被打印在鸟身上，每一个都链接到不同的网站。当用户扫码时，他们被传送到另一个现实中：一个充满视频、音频体验的现实中。

Night View 夜景

Details细节

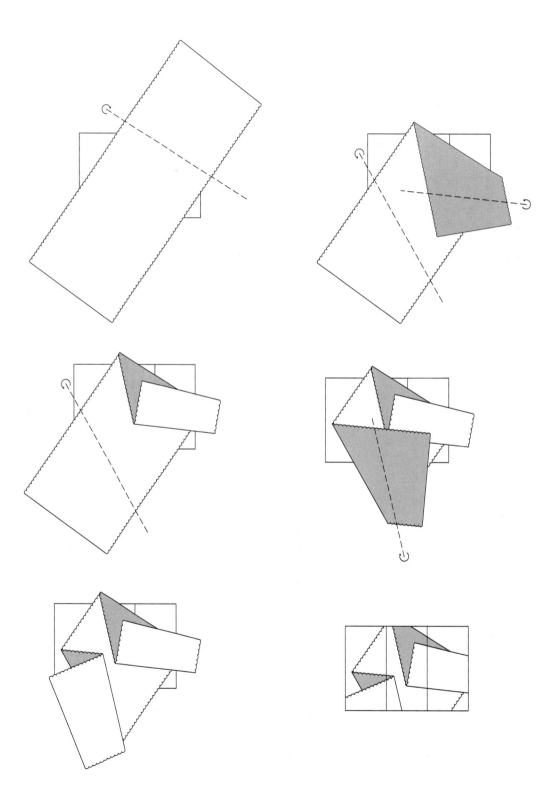

Entry Gate 入口门

Creating an emotive aesthetic experience is dealing with a product of an emotional input that will lead up to an emotional environment. During the design process, the architect's eye is constantly judging, sifting, and selecting the material in progress. It is about a constant flow of rational and irrational thinking, with no regards as to what comes first or last. Every designer involved has a visual talent, therefore, the images themselves are beautiful and seductive.

Elena Manferdini became involved with cultural preferences, with a personal insight, an individual articulation—as can be witnessed, for example, in her project, La Peer Hotel Gate, 2017—thus she affects the onlooker, the beholder, the participant, or the inhabitant—those it was designed for—similarly to music or film.

Yael Reisner

创造一种情绪化的审美体验是对一些情感数据的处理，它将引发一个有情绪的环境。在设计过程中，建筑师的眼睛一直在判断、筛查和选择设计过程中的物料。它是关于理性和非理性思维的持续循环，不考虑两种思维的出现顺序，每一个参与的设计师都有视觉天赋，因此，图像本身变得美丽而诱人。

埃琳娜·曼费迪尼成为文化偏好的一部分，以其个人的洞察力和表现力，如在"拉佩尔酒店大门"设计中体现出的那般，影响了她设计的受众：旁观者、观赏者、参与者或居民，就像音乐或电影所产生的影响一样。

雅尔·雷斯纳

Details细节

WOVEN HOUSE

INFONAVIT

Date: 2016
Client: Infonavit
Location: Tula de Allende, Mexico

编织房屋

因福那维公司

时间：2016年
客户：因福那维公司
地点：墨西哥，图拉—德阿连德市

"Woven House" is a proposal for a six unit condominium in the city of Tula de Allende in Hidalgo, Mexico. Tula de Allende has a rich history of ancient cultures dating back to the Toltec civilization. Reverberations of this history can be found from the scale of the city down to the textiles which are produced in the region. Designed as part of a competition hosted by Infonavit to create thoughtful low-income housing, "Woven House" also works to engage with the local context on multiple scales. The base plinth of the building which houses the common spaces is oriented to the city grid. As the building steps up to the more private spaces, the volume shifts to allow for a multidirectional view of the object. Referencing the context on a smaller scale, the building would be constructed from traditional brick methods in a pattern that references local weaving. With only two colors, "Woven House" is able to harness these traditional methods of brick-making and patterning to create a façade which has a high degree of visual complexity. The larger volume shifts of the building work closely with this patterning: the shifts in the building strategically allow for placement of windows while allowing the pattern to read as continuous in elevation.

"编织房屋"是墨西哥伊达尔戈州图拉—德阿连德市一套六单元公寓的提案。图拉—德阿连德有着丰富悠久的历史文化，可以追溯到托尔特克文明。从城市的规模到该地区生产的纺织品，都可以看到这段历史的印记。作为由因福那维公司主办的一项旨在创造精心设计的低收入者住房的比赛的一部分，"编织房屋"还致力于在多个层面上与当地环境进行互动。公共空间所在建筑的基座面向其城市整体脉络。随着建筑向更私密空间的深入，其主体会朝允许多角度观赏建筑本体的方向发生变化。考虑到较小尺度的环境，建筑将采用传统的砖砌方法，并参照当地的典型编织图案进行设计。伴随着仅有的两种主色调，"编织房屋"利用传统的制砖方法和图案，创造出一个具有高度视觉复杂性的外观。建筑的体量变化与这种图案密切相关：建筑的形态转换策略为窗户的放置创造了各种可能，同时允许图案在立面上连续显示。

Axon Drawing 轴测图

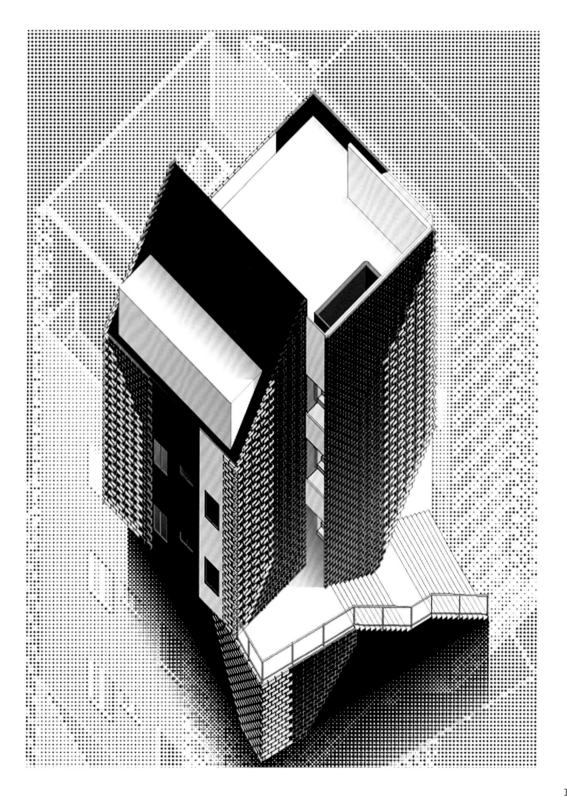

Physical Model 物理模型

Axon Drawing 轴测图

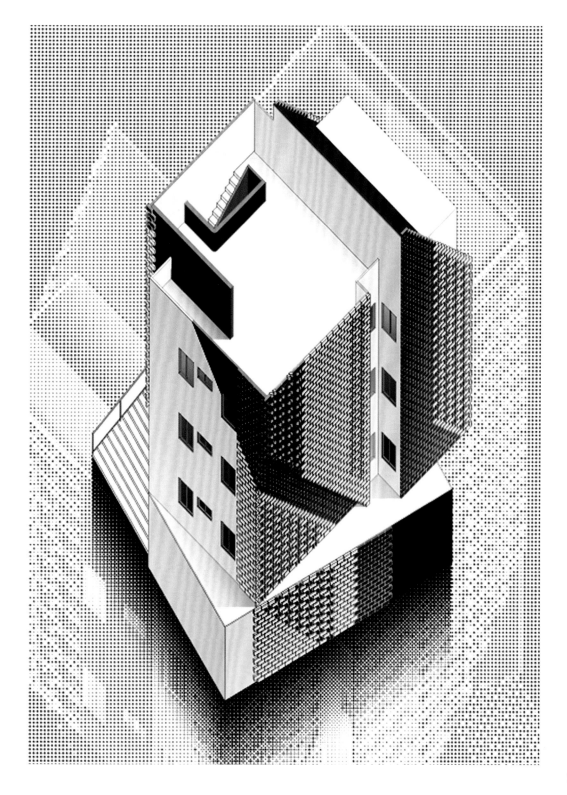

Elevation 立面

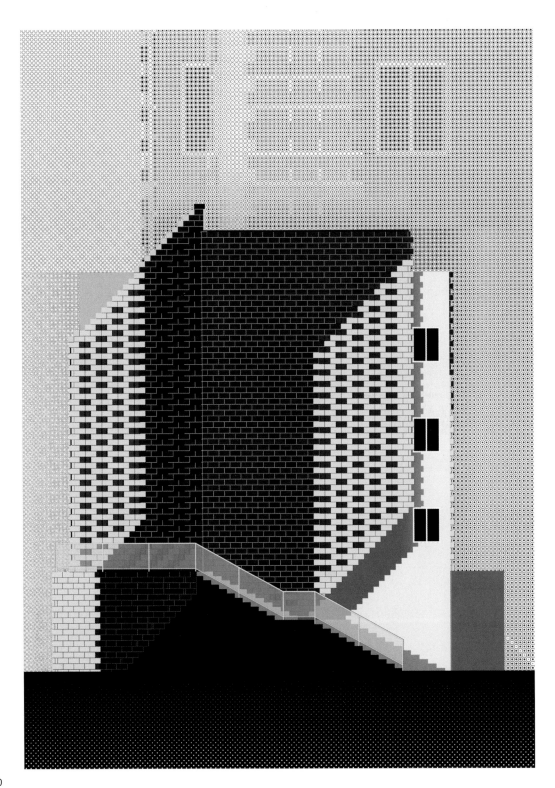

Elevation 立面

161

Elevation 立面

Elevation 立面

Elena Manferdini's projects deal with issues of 2D, 2.5D and 3D. Her work has a complicated ontology. It stands between a drawing and a surface object. I was invited to be a jury member for her studio and most of the projects argued this type of drawing discourse. After her drawing investigations, several colleagues were influenced and started to talk about them in this way. She has influenced a new way of thinking about contemporary drawings.

Gabriel Esquivel

埃琳娜·曼费迪尼的作品探讨了处理2D、2.5D和3D的问题。她的作品具有复杂的本体特性，介于图画作品和表面装饰之间。我受邀担任她工作室的评委会委员，现其大多数工程都在讨论这种类型的绘画语言。在她发表这一系列作品之后，几位同事受到了她作品的影响并开始以她的创作方式来讨论那些画作。对当代绘画，她开启了一种全新的思考方式。

加布里埃尔·埃斯奎维尔

Physical Model 物理模型

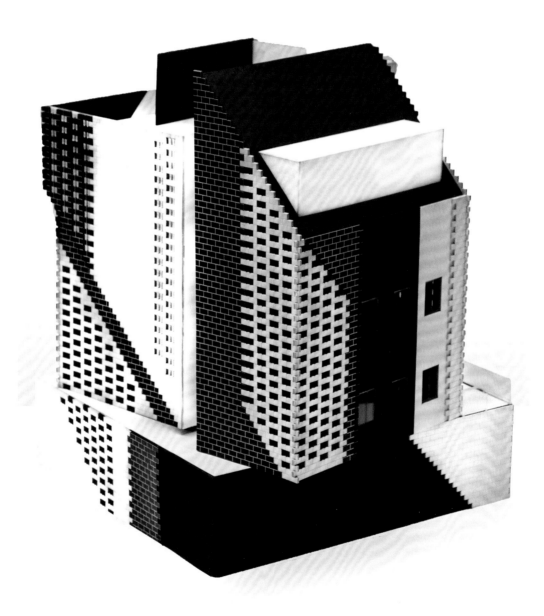

BIOGRAPHY

ELENA MANFERDINI

Elena Manferdini, principal and owner of Atelier Manferdini, has over fifteen years of professional experience in architecture, art, design, and education. She graduated from the University of Civil Engineering (Bologna, Italy) and later received her Master of Architecture and Urban Design from the University of California Los Angeles (Los Angeles, CA).

In 2004, she founded Atelier Manferdini in Venice, CA. The office has completed projects in the US, Europe, and Asia.

Atelier Manferdini is currently working on several ground-up multifamily housing projects now under construction in West Hollywood and San Fernando Valley. In 2018, Atelier Manferdini completed an 8 story façade for the Hermitage parking structure in Florida and the entry gate for La Peer Hotel in West Hollywood. In 2017, her firm was awarded the interior artwork for the Bethesda station of the Purple Line in Washington and she completed construction of the Alexander Montessori School façades in Miami. Notable among the firm's projects are the permanent public art projects for the interior lobby and outdoor open space of the Zev Yaroslavsky San Fernando Valley Family Support Center of LA County; the entryway for the Hubert Humphrey Hospital in Downtown Los Angeles and the remodel of a 3 floor cruise boat in Kyoto (Japan).

In addition to art and architectural projects, Elena Manferdini has worked on several museum installations and object designs. Through the years she has collaborated with internationally renowned companies such as: Lexus, BMW, Swarovski, Sephora, Driade, MTV, Fiat, Nike, Alessi, Ottaviani, Moroso, Valentino, Arktura, Lerival and TWG.

Elena Manferdini's work has appeared internationally in books, professional journals and reviews. She has been featured in several publications: Domus, New York Times, Elle, Vogue, ID, Icon, Form, Contemporary, Metropolis, and Architectural Design are selected examples.

In 2017, her façade for Alexander Montessori School in Miami, Florida received 2 AIA Miami design awards: People's Choice Award of Excellence and Merit Award of Excellence. Her artwork received the esteemed 2016 Public Art North America award. In addition, Elena Manferdini was awarded the 2013 COLA Fellowship given by City of Los Angeles Department of Cultural Affairs to support the production of original artwork. In 2013, she received a Graham Award for architecture, the 2013 ACADIA Innovative Research Award of Excellence. In 2011, she was one of the recipients of the prestigious annual grants from the United States Artists (USA) in the category of architecture and design. Finally, her Blossom design for Alessi received the Good Design Award in 2011.

Since 2003, Elena has been part of the design faculty at the Southern California Institute of Architecture (SCI-Arc), where she is now the Graduate Programs Chair. Throughout the years, Elena has been acknowledged with prestigious appointments such as the Howard Friedman Visiting Professor of Practice at the University of California Berkeley (UCB) and various Visiting Professor Positions at Cornell University, University of Pennsylvania, and Seika University. In 2013 she was selected as recipient for the Educator of the Year presidential award given by the AIA Los Angeles.

作者介绍

埃琳娜·曼费迪尼

埃琳娜·曼费迪尼是曼费迪尼工作室的主要负责人，在建筑、艺术、设计和教育方面拥有超过15年的专业经验。她毕业于意大利博洛尼亚的土木工程大学，随后在加利福尼亚大学洛杉矶分校（美国加利福尼亚州洛杉矶）获得了建筑和城市设计硕士学位。

2004年，她在加利福尼亚州威尼斯郡成立了曼费迪尼工作室，该工作室已在美国以及欧洲和亚洲的许多地区完成了很多项目。

曼费迪尼工作室目前正在西好莱坞和圣费尔南多谷进行数个正在建设中的住宅项目。2018年，曼费迪尼工作室完成了佛罗里达州的冬宫停车楼项目和西好莱坞拉佩尔酒店入口大门的8层外观设计。2017年，她的事务所的华盛顿地铁紫线贝塞斯达站设计被授予了室内艺术品奖，同年在迈阿密完成了亚历山大蒙特梭利学校外墙的建设。该公司的诸多项目中，洛杉矶县泽夫雅罗斯拉夫斯基圣费尔南多谷家庭支持中心的室内大厅和室外开放空间的永久公共艺术项目，洛杉矶市中心休伯特汉弗莱医院的入口通道，以及京都（日本）3层游轮的改造都在业内外产生了深远影响。

除了艺术和建筑项目，埃琳娜·曼费迪尼还参与了多个博物馆装置和静物设计。多年来，她一直与国际知名公司合作，如雷克萨斯、宝马、施华洛世奇、丝芙兰、德瑞德、纽约音乐电视、菲亚特、耐克、阿莱西、奥塔维亚尼、莫罗索、华伦天奴、阿尔图拉、拉瑞佛以及特威格。

埃琳娜·曼费迪尼的作品被刊登在各种国际化的书籍、专业期刊和艺术评论中。她曾在多个出版物中被大篇幅报道，如《多姆斯》《纽约时报》《建筑设计》等。

2017年，她在佛罗里达州迈阿密亚历山大蒙特梭利学校的立面设计获得了两项美国建筑师协会（AIA）迈阿密设计奖，分别是最受民众喜爱奖和卓越奖。她的作品获得了万众瞩目的2016年北美公共艺术奖。此外，埃琳娜·曼费迪尼还获得了洛杉矶市文化厅颁发的2013年COLA大奖，该奖项旨在支持原创艺术品的创作。2013年，她获得了格雷厄姆建筑奖，阿卡迪亚创新研究杰出奖。2011年，她参与美国艺术家在建筑和设计领域享有盛誉的年度奖金的角逐。还有，她为阿莱西设计的花朵在2011年获得了好设计奖。

自2003年以来，埃琳娜一直是南加州建筑学院设计系的一员，她现在是该学院的研究生院主任。多年来，埃琳娜一直被任命为加利福尼亚大学伯克利分校的霍华德·弗里德曼实践客座教授，以及康奈尔大学、宾夕法尼亚大学和塞卡大学等多个高校的客座教授，2013年，她被授予了洛杉矶美国建筑师协会（AIA）颁发的年度教育家总统奖。

CONTRIBUTORS
贡献者

Hernan Diaz Alonso

Hernan Diaz Alonso was born in Buenos Aires, Argentina in 1969, and holds a Master's in Advanced Architecture Design from the Graduate School of Architecture, Planning and Preservation (GSAPP) at Columbia University in New York City and a Bachelor of Architecture from the National University of Rosario, Argentina.

Before moving to Los Angeles, Hernan Diaz Alonso worked for architect Peter Eisenman as a senior designer with Eisenman Architects in New York City. After founding Xefirotarch in Los Angeles in 2001, today he currently serves as Principle of his new firm HDA-X Creative Agency. Hernan Diaz Alonso began teaching at SCI-Arc in Los Angeles in 2001 as design studio faculty, and served as Coordinator of the Graduate Thesis program from 2007–10, and Graduate Programs Chair from 2010-15. He was appointed SCI-Arc Director and CEO in September 2015.

埃尔南·迪亚兹·阿隆索

埃尔南·迪亚兹·阿隆索1969年出生于阿根廷布宜诺斯艾利斯，拥有纽约哥伦比亚大学建筑、规划与保护研究生院高级建筑设计硕士学位，以及阿根廷罗萨里奥国立大学建筑学学士学位。

埃尔南·迪亚兹·阿隆索曾在纽约艾森曼建筑师事务所担任高级设计师。后搬到洛杉矶，于2001年在洛杉矶创立了谢菲罗塔赫事务所。现在是新公司HDA-X创意机构的负责人。埃尔南·迪亚兹·阿隆索于2001年开始在洛杉矶的南加州建筑学院担任设计课程教师，2007年至2010年担任研究生论文课程的辅导员，并于2010年至2015年担任研究生院主任。他于2015年9月被任命为南加州建筑学院董事兼首席执行官。

Jasmine Benyamin

Jasmine Benyamin is Associate Professor of Architecture at the University of Wisconsin-Milwaukee, where she teaches history and theory courses as well as design studios at the undergraduate and graduate levels.

She hold bachelor's degrees in Architecture and French Literature from Columbia University, a Master of Architecture degree from Yale University, and a PhD from Princeton University. Her dissertation research was the recipient of several grants, including the Canadian Centre for Architecture (CCA) and fellowships from both the DAAD and Fulbright Foundation. Her publications include essays and reviews for the *JSAH, Thresholds, Constructs,* the *JAE, Offramp,* and the *Journal of Architecture*.

杰思敏·本雅明

杰思敏·本雅明是威斯康星大学密尔沃基分校的建筑学副教授，教授历史和理论课程，并主理本科和研究生水平的设计工作室。

她拥有哥伦比亚大学建筑与法国文学学士学位，耶鲁大学建筑硕士学位，普林斯顿大学博士学位。她的论文研究得到多项资助，包括加拿大建筑中心以及来自德国学术交流中心和富布赖特基金会的奖学金。她的出版物包括为《JSAH》《JAE》《Offramp》和《建筑杂志》等所著的文章和评论。

Gabriel Esquivel

Born and educated as an architect in Mexico City, with a Master's Degree from Ohio State University, Gabriel Esquivel is the director of the T4T (Technology and Theory) Lab, where the mission is to invite young practitioners from all over the world to develop their ideas within a lab/studio format.

Developing his early professional career at the office of NBBJ, he has completed projects across the globe that range across architectural typologies. After joining the architecture faculty at Texas A&M University, he became a promoter of new ideas in architecture. Gabriel is interested in geometry, representation, aesthetics and material logic and his work examines the integration of digital techniques and analogue conventions and its connections to architectural theory.

Jeffrey Kipnis

Jeffrey Kipnis is a professor of architecture at the Knowlton School where he teaches courses on architectural design and theory. For more than three decades, Kipnis' work has shaped the thinking, imagination and creative work of architects and critics. From seminal studies of the work of such key practitioners as Philip Johnson, Peter Eisenman, Rem Koolhaas and Daniel Libeskind, to theoretical reflections on the intellectual, cultural and political role of contemporary architecture in such essays as "Toward a New Architecture," "Twisting the Separatrix" and "Political Space I," to exhibitions on architectural drawing and design, Kipnis has brought a restless, generous and provocative originality to bear on the issues that have defined contemporary architecture.

Kipnis' writings on art and architecture have appeared in such publications as *Log, Hunch, Harvard Design Magazine, Quaderns, 2G, El Croquis, Art Forum, Assemblage*, and his books include *Choral Works: The Eisenman-Derrida Collaboration, Perfect Acts of Architecture*, and *The Glass House*.

加布里埃尔·埃斯奎维尔

加布里埃尔·埃斯奎维尔是墨西哥城的一名建筑师，拥有俄亥俄州立大学的硕士学位，是T4T（技术与理论）实验室的主任，该实验室的主要工作是邀请来自世界各地的年轻从业者以工作室的形式实现他们的想法。

他在NBBJ建筑师事务所开始了自己早期的职业生涯，在全球范围内完成了各种类型的建筑项目。在加入得克萨斯州农工大学建筑系之后，他成了建筑新理念的推动者。加布里埃尔对几何学、表现、美学和材料逻辑感兴趣，他的作品研究了数字技术和传统模拟的融合及其与建筑理论的联系。

杰弗雷·基普尼斯

杰弗雷·基普尼斯是诺尔顿学院的建筑学教授，教授建筑设计和理论课程。30多年来，基普尼斯塑造了兼具建筑师和批评家思维的，富有想象和创造力的作品。从对菲利普·约翰逊、彼得·艾森曼、雷姆·库哈斯和丹尼尔·利伯斯金等重要行业实践者的作品的开创性研究，到在"走向新建筑""扭曲分离主义"和"政治空间I"等文章中对当代建筑的知识、文化和政治作用的理论思考，再到对建筑绘图和设计的展现，基普尼斯秉承了一种永不停歇的、真诚而富有挑战的原创性，来定义当代建筑的问题。

基普尼斯关于艺术和建筑的文章已经出现在诸如《Log》《Hunch》《哈佛设计杂志》《Quaderns》《2G》《El Croquis》《Art Forum》《Assemblage》等杂志中，他的著作包括《艾森曼和德里达的合作》《完美的建筑行为》和《玻璃屋》。

LI NING

Li Ning, born in 1982, received his Ph.D. from Tsinghua University in Beijing. National Class 1 Registered Architect in China, he is the Deputy Director of the Department of Architecture in Beijing University of Technology, Deputy Chief Architect of Company of Architecture Design of Beijing University of Technology. Ning's expertise lies in non-linear architectural design; parametric design, shape generation by algorithms, performance simulation, and data mining. He is the author of *Digital Diagram from BIO-Form for Architectural Design*, co-author Xu Weiguo, and been published with thirteen papers.

Li Ning's completed works include; Zhuhai Garden of Nanning World Expo 2018, the renovation, construction and landscape design of the Zhongguancun Industrial Park, the construction of Atrium of No. 1Kaidi Building in Wuhan, and the campus planning and individual design of Anyang Normal University.

XU WEIGUO

Xu Weiguo is professor and chair of the Architecture Department in the school of architecture at Tsinghua University. He was a visiting scholar at MIT in 2007 and taught in SCI-Arc and USC in 2011-2012. He studied architecture at Tsinghua University, and began teaching at the same institution before moving to Japan to work for Murano Mori Architects. He was awarded his doctorate from Kyoto University in Japan. Upon returning to China, he established his own architectural practice (XWG) in Beijing. He has been the recipient of many awards, and his 70 works have been published in an array of journals. He is also the author of 11 books.

Xu Weiguo was included in the Exhibition of Young Chinese Architects at the XX International UIA Congress in 1999, and was selected as one of the architects to represent China in the A1 pavilion at ABB2004. He was one of the curators of the China International Architectural Biennial in Beijing in 2004, 2006 2008, and 2010.

李宁

李宁，1982年出生，于清华大学获博士学位。中国国家一级注册建筑师，北京工业大学建筑系副主任、北京工业大学建筑与勘察设计院副总建筑师。研究领域为非线性建筑设计、参数化设计、算法生形、性能模拟、数据挖掘。著有《生物形态的建筑数字图解》，发表论文十余篇。

李宁的已建成作品包括2018年南宁世博珠海园，中关村工业园区建筑与景观改造设计，武汉凯迪一号楼中庭加建，安阳师范大学校园规划与单体设计等多项工程。

徐卫国

徐卫国是清华大学建筑学院建筑系教授兼系主任。他于2007年在麻省理工学院担任访问学者，并于2011年至2012年在南加州建筑学院和南加州大学任教。他在清华攻读建筑学学位，并留校任教。随后前往日本在村野森建筑事务所工作。在取得日本京都大学博士学位后，返回中国，在北京成立了个人建筑设计（XWG）工作室。他是许多奖项的获得者，其70余件作品已在一系列期刊上发表，同时著有11本专著。

徐卫国于1999年参加世界建筑师大会（UIA）的中国青年建筑师作品展，并被选为代表中国在2004年首届中国国际建筑艺术双年展（ABB2004)的A1展馆参展的建筑师之一。他是2004年、2006年、2008年和2010年在北京举办的中国国际建筑艺术双年展的策展人之一。

YAEL REISNER

Dr Reisner, an architect registered in Israel, born in Tel Aviv, and living in London since 1990; a designer, researcher, writer and curator. RMIT Melbourne's PhD, AA Diploma, RIBA part 1 & 2, BSc in Biology.

Yael Reisner Studio is an architectural research led practice.

Since 1997 she taught architectural design, nine years at the Bartlett, ten years internationally. In 2017 she was guest professor at PBSA-HSD Düsseldorf. Her book, *Yael Reisner with Fleur Watson, Architecture and Beauty, Conversations with Architects about A Troubled Relationship*, Wiley, 2010, translated to Chinese in 2014.

Reisner's installation, Take My Hand, Rights and Weddings, Barcelona, 2014, designed and managed at her studio.

She is guest-editor of September 2019's Issue of Wiley's *AD magazine*, entitled: *Beauty Matters; human judgment and the pursuit of new beauties in post-digital architecture*. Reisner is the head curator of the Tallinn Architecture Biennale, September 2019, entitled: Beauty Matters; the Resurgence of Beauty.

雅尔·雷斯纳

雷斯纳博士，以色列注册建筑师，生于特拉维夫，1990年起居住在伦敦；设计师、研究员、作家和策展人。墨尔本皇家理工学院博士，英国建筑同盟（AA）毕业生，英国皇家建筑师协会成员，生物学、理学学士。

雅尔·雷斯纳工作室是一家以建筑研究为主导的公司。

从1997年开始，雅尔·雷斯纳开始教授建筑设计，从教于巴特利特9年，之后辗转于国际各地从教10年。2017年，她成为杜塞尔多夫高等专业学院的客座教授。她的书：《雅尔·雷斯纳与弗勒尔·沃森》《建筑与美》《与建筑师讨论困境关系》于2010年威利出版社出版，并于2014年翻译成中文。

雷斯纳的装置作品，《牵着我的手》《权力和婚礼》，于2014年在她巴塞罗那的工作室完成设计，并由工作室管理。

她是威利出版社《建筑设计》杂志2019年9月期的客座编辑，著文题为《美很重要：人类的判断力和后数字化建筑中对全新美学的追求》。雷斯纳是2019年9月塔林建筑双年展的首席策展人，主题为美很重要，美的复苏。

IMAGE ANNOTATION
图片注释

BUILDING THE PICTURE 构建图片

009　Building the Picture Ⅳ | Art Institute of Chicago | 2015 | Print on Powder Coated Aluminum | 60 cm × 60 cm
　　构建图片Ⅳ | 芝加哥艺术博物馆 | 2015年 | 在粉末涂层铝上打印 | 60 cm × 60 cm

010　Building the Picture Ⅲ | Art Institute of Chicago | 2015 | Print on Powder Coated Aluminum | 60 cm × 60 cm
　　构建图片Ⅲ | 芝加哥艺术博物馆 | 2015年 | 在粉末涂层铝上打印 | 60 cm × 60 cm

011　Building the Picture Ⅴ | Art Institute of Chicago | 2015 | Print on Powder Coated Aluminum | 60 cm × 60 cm
　　构建图片 Ⅴ | 芝加哥艺术博物馆 | 2015年 | 在粉末涂层铝上打印 | 60 cm × 60 cm

012　Grids and Typologies Ⅳ | Art Institute of Chicago | 2015 | Archival Ink Prints | 48 cm × 33 cm
　　网格和类型Ⅳ | 芝加哥艺术博物馆 | 2015年 | 档案墨水打印 | 48 cm × 33 cm

015　Grids and Typologies Ⅲ | Art Institute of Chicago | 2015 | Archival Ink Prints | 33 cm × 48 cm
　　网格和类型Ⅲ | 芝加哥艺术博物馆 | 2015年 | 档案墨水打印 | 33 cm × 48 cm

016　Building the Picture Ⅵ | Art Institute of Chicago | 2015 | Print on Powder Coated Aluminum | 60 cm × 60 cm
　　构建图片Ⅵ | 芝加哥艺术博物馆 | 2015年 | 在粉末涂层铝上打印 | 60 cm × 60 cm

017　Building the Picture Ⅷ | Art Institute of Chicago | 2015 | Print on Powder Coated Aluminum | 60 cm × 60 cm
　　构建图片Ⅷ | 芝加哥艺术博物馆 | 2015年 | 在粉末涂层铝上打印 | 60 cm × 60 cm

019　Grids and Typologies Ⅱ | Art Institute of Chicago | 2015 | Archival Ink Prints | 33 cm × 48 cm
　　网格和类型Ⅱ | 芝加哥艺术博物馆 | 2015年 | 档案墨水打印 | 33 cm × 48 cm

020　Grids and Typologies I | Art Institute of Chicago | 2015 | Archival Ink Prints | 48 cm × 33 cm
　　网格和类型I | 芝加哥艺术博物馆 | 2015年 | 档案墨水打印 | 48 cm × 33 cm

023　Building the Picture I | Art Institute of Chicago | 2015 | Print on Powder Coated Aluminium | 60cm × 122cm
　　构建图片I | 芝加哥艺术博物馆 | 2015年 | 在粉末涂层铝上打印 | 60cm × 122cm

024　Building the Picture Ⅹ | Art Institute of Chicago | 2015 | Print on Powder Coated Aluminum | 60 cm × 60 cm
　　构建图片Ⅹ | 芝加哥艺术博物馆 | 2015年 | 在粉末涂层铝上打印 | 60 cm × 60 cm

025　Building the Picture ⅩⅢ | Art Institute of Chicago | 2015 | Print on Powder Coated Aluminum | 60 cm × 60 cm
　　构建图片ⅩⅢ | 芝加哥艺术博物馆 | 2015年 | 在粉末涂层铝上打印 | 60 cm × 60 cm

026　Building the Picture Ⅸ | Art Institute of Chicago | 2015 | Print on Powder Coated Aluminum | 60 cm × 60 cm
　　构建图片Ⅸ | 芝加哥艺术博物馆 | 2015年 | 在粉末涂层铝上打印 | 60 cm × 60 cm

027　Building the Picture ⅩI | Art Institute of Chicago | 2015 | Print on Powder Coated Aluminum | 60 cm × 60 cm
　　构建图片ⅩI | 芝加哥艺术博物馆 | 2015年 | 在粉末涂层铝上打印 | 60 cm × 60 cm

028　Building the Picture – Unfold I | Art Institute of Chicago | 2015 | Archival Ink Prints | 33 cm × 48 cm
　　物理模型的安排展开I | 芝加哥艺术学院 | 2015年 | 档案墨水打印 | 33 cm × 48 cm

030　Building the Picture – Unfold Ⅱ | Art Institute of Chicago | 2015 | Archival Ink Prints | 33 cm × 48 cm
　　物理模型的安排展开Ⅱ | 芝加哥艺术学院 | 2015年 | 档案墨水打印 | 33 cm × 48 cm

032 Building the Picture – Unfold Ⅲ | Art Institute of Chicago | 2015 | Archival Ink Prints | 33 cm × 48 cm
物理模型的安排展开Ⅲ | 芝加哥艺术学院 | 2015年 | 档案墨水打印 | 33 cm × 48 cm

034 Fictional Chicago Skyline | Art Institute of Chicago | 2015
虚构的芝加哥天际线 | 芝加哥艺术学院 | 2015年

036 Fictional Chicago Skyline | Art Institute of Chicago | 2015
虚构的芝加哥天际线 | 芝加哥艺术学院 | 2015年

BUILDING PORTRAITS 建筑肖像

039 Building Portraits Ⅴ | Industry Gallery | 2015 | Archival Ink Prints | 33 cm × 48 cm
建筑肖像Ⅴ | 工业画廊 | 2015年 | 档案墨水打印 | 33 cm × 48 cm

041 Building Portraits Ⅵ | Industry Gallery | 2015 | Archival Ink Prints | 33 cm × 48 cm
建筑肖像Ⅵ | 工业画廊 | 2015年 | 档案墨水打印 | 33 cm × 48 cm

042 Building Portraits Ⅶ | Industry Gallery | 2015 | Archival Ink Prints | 33 cm × 48 cm
建筑肖像Ⅶ | 工业画廊 | 2015年 | 档案墨水打印 | 33 cm × 48 cm

043 Building Portraits Ⅷ | Industry Gallery | 2015 | Archival Ink Prints | 33 cm × 48 cm
建筑肖像Ⅷ | 工业画廊 | 2015年 | 档案墨水打印 | 33 cm × 48 cm

044 Fire Station – Ⅰ to Ⅳ | Industry Gallery | 2015 | Archival Ink Prints | 33 cm × 48 cm
消防局—Ⅰ到Ⅳ | 工业画廊 | 2015年 | 档案墨水打印 | 33 cm × 48 cm

046 Postures and Ground Ⅴ | Industry Gallery | 2015 | Archival Ink Prints | 33 cm × 48 cm
姿势和地面Ⅴ | 工业画廊 | 2015年 | 档案墨水打印 | 33 cm × 48 cm

047 Postures and Ground Ⅵ | Industry Gallery | 2015 | Archival Ink Prints | 33 cm × 48 cm
姿势和地面Ⅵ | 工业画廊 | 2015年 | 档案墨水打印 | 33 cm × 48 cm

049 Postures and Ground Ⅰ | Industry Gallery | 2015 | Archival Ink Prints | 33 cm × 48 cm
姿势和地面Ⅰ | 工业画廊 | 2015年 | 档案墨水打印 | 33 cm × 48 cm

050 Building Portraits Ⅳ | Industry Gallery | 2015 | Archival Ink Prints | 48 cm × 33 cm
建筑肖像Ⅳ | 工业画廊 | 2015年 | 档案墨水打印 | 48 cm × 33 cm

053 Shapes and Ground Ⅲ | Industry Gallery | 2015 | Archival Ink Prints | 33 cm × 48 cm
形状和地面Ⅲ | 工业画廊 | 2015年 | 档案墨水打印 | 33 cm × 48 cm

054 Shapes and Ground Ⅰ | Industry Gallery | 2015 | Archival Ink Prints | 33 cm × 48 cm
形状和地面Ⅰ | 工业画廊 | 2015年 | 档案墨水打印 | 33 cm × 48 cm

055 Shapes and Ground Ⅱ | Industry Gallery | 2015 | Archival Ink Prints | 33 cm × 48 cm
形状和地面Ⅱ | 工业画廊 | 2015年 | 档案墨水打印 | 33 cm × 48 cm

056 Building Portraits Ⅲ | Industry Gallery | 2015 | Archival Ink Prints | 48 cm × 33 cm
建筑肖像Ⅲ | 工业画廊 | 2015年 | 档案墨水打印 | 48 cm × 33 cm

INK ON MIRROR 镜子上的墨水

059 Folds & Pleats Ⅱ | Pacific Design Center | 2016 | Print on Acrylic | 60 cm × 92 cm
折叠和褶皱Ⅱ | 太平洋设计中心 | 2016年 | 在亚克力上打印 | 60 cm × 92 cm

060　Folds & Pleats – Ⅲ and I | Pacific Design Center | 2016 | Print on Acrylic | 60 cm × 92 cm
　　　折叠和褶皱 – Ⅲ和I | 太平洋设计中心 | 2016年 | 在亚克力上打印 | 60 cm × 92 cm

063　Building Portraits V – Model Ⅱ | Pacific Design Center | 2016 | Wood Plinth and Acrylic Physical Model
　　　建筑肖像V – 模型Ⅱ | 太平洋设计中心 | 2016年 | 木质底座和亚克力物理模型

064　Postures and Ground Ⅱ | Pacific Design Center | 2016 | Archival Print | 33 cm × 48 cm
　　　姿势和地面Ⅱ | 太平洋设计中心 | 2016年 | 档案打印 | 33 cm × 48 cm

065　Postures and Ground Ⅱ Model | Pacific Design Center | 2016 | Wood Plinth and Acrylic Physical Model
　　　姿势和地面Ⅱ模型 | 太平洋设计中心 | 2016年 | 木质底座和亚克力物理模型

066　Postures and Ground Ⅱ Unroll Drawing | Pacific Design Center | 2016 | Archival Ink Print | 48 cm × 33 cm
　　　姿势和地面Ⅱ展开绘图 | 太平洋设计中心 | 2016年 | 档案墨水打印 | 48 cm × 33 cm

068　Postures and Ground Ⅲ | Pacific Design Center | 2016 | Archival Print | 33 cm × 48 cm
　　　姿势和地面Ⅲ | 太平洋设计中心 | 2016年 | 档案打印 | 33 cm × 48 cm

069　Postures and Ground Ⅲ Model | Pacific Design Center | 2016 | Wood Plinth and Acrylic Physical Model
　　　姿势和地面Ⅲ模型 | 太平洋设计中心 | 2016年 | 木质底座和亚克力物理模型

070　Postures and Ground Ⅲ Unroll Drawing | Pacific Design Center | 2016 | Archival Ink Print | 48 cm × 33 cm
　　　姿势和地面Ⅲ展开绘图 | 太平洋设计中心 | 2016年 | 档案墨水打印 | 48 cm × 33 cm

072　Forms and Ground – Ⅰ to Ⅲ | Pacific Design Center | 2016 | Archival Ink Prints | 33 cm × 48 cm
　　　表格和地面 – Ⅰ至Ⅲ | 太平洋设计中心 | 2016年 | 档案墨水打印 | 33 cm × 48 cm

074　Forms and Ground Ⅰ – Ⅲ Models | Pacific Design Center | 2016 | Wood Plinth and Acrylic Physical Models
　　　表格和地面Ⅰ – Ⅲ模型 | 太平洋设计中心 | 2016年 | 木质底座和亚克力物理模型

076　Forms and Ground Ⅱ Unroll Drawing | Pacific Design Center | 2016 | Archival Ink Print | 48 cm × 33 cm
　　　表格和地面Ⅱ展开绘图 | 太平洋设计中心 | 2016年 | 档案墨水打印 | 48 cm × 33 cm

077　Forms and Ground I Unroll Drawing | Pacific Design Center | 2016 | Archival Ink Print | 48 cm × 33 cm
　　　表格和地面Ⅰ展开绘图 | 太平洋设计中心 | 2016年 | 档案墨水打印 | 48 cm × 33 cm

078　Postures and Ground Ⅳ | Pacific Design Center | 2016年 | Archival Ink Prints | 33 cm × 48 cm
　　　姿势和地面Ⅳ | 太平洋设计中心 | 2016年 | 档案墨水打印 | 33cm x48cm

079　Postures and Ground Ⅳ – Model | Pacific Design Center | 2016 | Wood Plinth and Acrylic Physical Model
　　　姿势和地面Ⅳ – 模型 | 太平洋设计中心 | 2016年 | 木质底座和亚克力物理模型

080　Postures and Ground Ⅳ Unroll Drawing | Pacific Design Center | 2016 | Archival Ink Print | 48 cm × 33 cm
　　　姿势和地面Ⅳ展开绘图 | 太平洋设计中心 | 2016年 | 档案墨水打印 | 48 cm × 33 cm

082　Postures and Ground V Unroll Drawing | Pacific Design Center | 2016 | Archival Ink Print | 48 cm × 33 cm
　　　姿势和地面V展开绘图 | 太平洋设计中心 | 2016年 | 档案墨水打印 | 48 cm × 33 cm

084　Postures and Ground Ⅳ – V Models | Pacific Design Center | 2016 | Wood Plinth and Acrylic Physical Models
　　　姿势和地面Ⅳ – V模型 | 太平洋设计中心 | 2016年 | 木质底座和亚克力物理模型

086　Forms and Ground Ⅲ Model | Pacific Design Center | 2016 | Wood Plinth and Acrylic Physical Model
　　　表格和地面Ⅲ模型 | 太平洋设计中心 | 2016年 | 木质底座和亚克力物理模型

087　Forms and Ground Ⅲ Unroll Drawing | Pacific Design Center | 2016 | Archival Ink Print | 48 cm × 33 cm
　　　表格和地面Ⅲ展开绘图 | 太平洋设计中心 | 2016年 | 档案墨水打印 | 48 cm × 33 cm

089　Forms and Ground V | Pacific Design Center | 2016 | Print on Powder Coated Steel, Gloss Coat | 60 cm × 92 cm
　　表格和地面V | 太平洋设计中心 | 2016年 | 在粉末涂层钢，光泽涂层上打印 | 60 cm × 92 cm

090　Forms and Ground Ⅵ | Pacific Design Center | 2016 | Print on Powder Coated Steel, Gloss Coat | 60 cm × 92 cm
　　表格和地面Ⅵ | 太平洋设计中心 | 2016年 | 在粉末涂层钢，光泽涂层上打印 | 60 cm × 92 cm

091　Forms and Ground Ⅳ | Pacific Design Center | 2016 | Print on Powder Coated Steel, Gloss Coat | 60 cm × 92 cm
　　表格和地面Ⅳ | 太平洋设计中心 | 2016年 | 在粉末涂层钢，光泽涂层上打印 | 60 cm × 92 cm

092　Forms and Ground Ⅳ Model | Pacific Design Center | 2016 | Wood Plinth and Acrylic Physical Model
　　表格和地面Ⅳ模型 | 太平洋设计中心 | 2016年 | 木质底座和亚克力物理模型

093　Forms and Ground Ⅳ Unroll Drawing | Pacific Design Center | 2016 | Archival ink | 48 cm × 33 cm
　　表格和地面Ⅳ展开绘图 | 太平洋设计中心 | 2016年 | 档案墨水 | 48 cm × 33 cm

094　Forms and Ground Ⅵ Model | Pacific Design Center | 2016 | Wood Plinth and Acrylic Physical Model
　　表格和地面Ⅵ模型 | 太平洋设计中心 | 2016年 | 木质底座和亚克力物理模型

095　Forms and Ground Ⅵ Unroll Drawing | Pacific Design Center | 2016 | Archival ink | 33 cm × 48 cm
　　表格和地面Ⅵ展开绘图 | 太平洋设计中心 | 2016年 | 档案墨水 | 33 cm × 48 cm

096　Suite of 9 Models | Pacific Design Center | 2016 | Wood Plinth and Acrylic Physical Model
　　9个套房模型 | 太平洋设计中心 | 2016年 | 木质底座和亚克力物理模型

TWONESS 两元一体

099　26_Doppler_21 | 18 cm × 13 cm | 16_Doppler_13 | 2016 | 18 cm × 13 cm
　　26_多普勒_21 | 18 cm × 13 cm | 16_多普勒_13 | 2016 | 18 cm × 13 cm

100　01_Doppler_01 | 18 cm × 13 cm | 02_Doppler_11 | 18 cm × 13 cm
　　01_多普勒_01 | 18 cm × 13 cm | 02_多普勒_11 | 18 cm × 13 cm

101　03_Doppler_02 | 18 cm × 13 cm | 04_Doppler_23 | 18 cm × 13 cm
　　03_多普勒_02 | 18 cm × 13 cm | 04_多普勒_23 | 18 cm × 13 cm

103　07_Doppler_17 | 18 cm × 13 cm | 08_Doppler_05 | 18 cm × 13 cm
　　07_多普勒_17 | 18 cm × 13 cm | 08_多普勒_05 | 18 cm × 13 cm

105　15_Doppler_27 | 18 cm × 13 cm | 13_Doppler_20 | 18 cm × 13 cm
　　15_多普勒_27 | 18 cm × 13 cm | 13_多普勒_20 | 18 cm × 13 cm

106　27_Doppler_25 | 18 cm × 13 cm | 28_Doppler_29 | 18 cm × 13 cm
　　27_多普勒_25 | 18 cm × 13 cm | 28_多普勒_29 | 18 cm × 13 cm

107　17_Doppler_14 | 18 cm × 13 cm | 20_Doppler_15 | 18 cm × 13 cm
　　17_多普勒_14 | 18 cm × 13 cm | 20_多普勒_15 | 18 cm × 13 cm

CHICAGO SKYLINE 芝加哥天际线

109　Matrix of Folded Facades | Chicago Architecture Biennial | 2017 | Prints on Acrylic
　　矩阵和折叠立面 | 芝加哥建筑双年展 | 2017年 | 丙烯酸打印

110　Elevations | Chicago Architecture Biennial | 2017 | Prints on Acrylic | 122 cm × 244 cm
　　立面 | 芝加哥建筑双年展 | 2017年 | 丙烯酸打印 | 122 cm × 244 cm

AYALA 阿亚拉

113　East Facade I West Hollywood CA I 2019
　　 东立面I加利福尼亚州西好莱坞市I2019年

114　North Facade I West Hollywood CA I 2019
　　 北立面I 加利福尼亚州西好莱坞市I2019年

116　West Facade I West Hollywood CA I 2019
　　 西立面I加利福尼亚州西好莱坞市I2019年

118　North Facade I West Hollywood CA I 2019
　　 北立面I加利福尼亚州西好莱坞市I2019年

120　East Elevation I West Hollywood CA I 2019
　　 东立面I加利福尼亚州西好莱坞市I2019年

121　West Elevation I West Hollywood CA I 2019
　　 西立面I加利福尼亚州西好莱坞市I2019年

122　South Elevation I West Hollywood CA I 2019
　　 南立面I加利福尼亚州西好莱坞市I2019年

124　North and South Elevations I West Hollywood CA I 2019
　　 南北立面I加利福尼亚州西好莱坞市I2019年

126　East Elevation I West Hollywood CA I 2019
　　 东立面I加利福尼亚州西好莱坞市I2019年

127　West Elevation I West Hollywood CA I 2019
　　 西立面I 加利福尼亚州西好莱坞市I2019年

AT HUMAN SCALE 人体尺度

129　Digital Print IV I BMW I 2015 I Archival Ink Print I 33 cm × 48 cm
　　 数码打印IV I 宝马I 2015年I 档案墨水打印I 33 cm × 48 cm

130　Digital Print II I BMW I 2015 I Archival Ink Print I33 cm × 48 cm
　　 数码打印II I 宝马I 2015年I 档案墨水打印I33 cm × 48 cm

131　Digital Print Ⅲ I BMW I 2015 I Archival Ink Print I 33 cm × 48 cm
　　 数码打印Ⅲ I 宝马I 2015年I 档案墨水打印I33 cm × 48 cm

132　Installation I BMW I 2015
　　 安装I 宝马I2015年

134　Installation I BMW I 2015
　　 安装I 宝马I2015年

BLANK FACADE 空白立面

137　South Elevation I Hermitage Garage I 2018
　　 南立面I 冬宫停车楼I2018年

138　Facade Detail I Hermitage Garage I 2018
　　 立面细节I 冬宫停车楼I2018年

ALEXANDER MONTESSORI SCHOOL 亚历山大蒙特梭利学校

141　Facade Shading Panels I Alexander Montessori School I 2016
　　 立面遮阳板I 亚历山大蒙特梭利学校I2016年

144　Night View I Alexander Montessori School I 2016
　　 夜景I 亚历山大蒙特梭利学校I2016年

146　Interior View I Alexander Montessori School I 2016
　　 内部视图I 亚历山大蒙特梭利学校I2016年

CABINET OF WONDERS 神奇陈列柜

149　Night View I La Peer Hotel I 2017
　　 夜景I 拉佩尔酒店 I2017年

150　Details I La Peer Hotel I 2017
　　 细节I 拉佩尔酒店 I2017年

152　Entry Gate I La Peer Hotel I 2017
　　 入口门I 拉佩尔酒店 I2017年

155　Details I La Peer Hotel I 2017
　　 细节I 拉佩尔酒店 I2017年

WOVEN HOUSE 编织房屋

157　Axon Drawing I Infonavit I 2016
　　 轴测图I 因福那维公司 I2016年

158　Physical Model I Infonavit I 2016
　　 物理模型I 因福那维公司 I2016年

159　Axon Drawing I Infonavit I 2016
　　 轴测图I 因福那维公司 I2016年

160　Elevation I Infonavit I 2016
　　 立面I 因福那维公司 I2016年

161　Elevation I Infonavit I 2016
　　 立面I 因福那维公司 I2016年

162　Elevation I Infonavit I 2016
　　 立面I 因福那维公司 I2016年

163　Elevation I Infonavit I 2016
　　 立面I 因福那维公司 I2016年

165　Physical Model I Infonavit I 2016
　　 物理模型I 因福那维公司 I2016年